To all who love our Forest Lan[...]

Howard [...]

I Remember Logging

1923-1997

By

Howard P. Brunson

Edited
by
Marjorie M. Watkins

Fir Tree Press
Vashon, Washington

I Remember Logging

Copyright 1998 © by Howard P. Brunson
Library of Congress Number 97-94756
ISBN 0-9661667-0-1
All rights reserved.

Published by
Fir Tree Press, a division of Celesticomp, Inc.

No part of this book may be reproduced without prior written permission of the publisher by any electronic or mechanical means, including information storage and retrieval systems, except by educational institutions for instruction only, and provided title and author of this book are mentioned. Also excepted are reviewers, who may quote brief passages of this book in reviews. For any other use, please request permission from the publisher. For information, contact: Fir Tree Press, 8903 SW Bayview Drive, Vashon, WA 98070. FAX: (206)463-6688.

Limited
First Edition

Printed in the United States of America
1998

DEDICATION

I dedicate this book to a better
understanding of the value and uses of
our forest lands.

ACKNOWLEDGEMENTS

I want to thank all the host of relatives and friends who have helped in many ways to make it possible to publish this book of my logging experiences. Several women helped with typing my handwriting, some people supplied pictures, others contributed money to help pay the publishing expense. I hesitate to write their names, as there were so many who helped without thought of payment. I will mention my daughter, Marjorie, without whose countless hours spent in editing, typesetting, and book design, this book would not have come out in its present form.

We also wish to acknowledge the courtesy and helpfulness of photo librarians at Oregon and Washington state historical societies, Oregon State Library, Tacoma Public Library, and the University of Washington Allen Library.

We all feel that this book contains essential history of the logging of our Oregon forests, telling how logging was done in the past, and how it is done now. It tells of the improvements in logging methods to stop waste, to make the working conditions better and safer for the loggers, and to manage our timberlands as a renewable resource, with regard for the animals and the ecology.

Howard Brunson

Cover Photo:
Howard Brunson and Norman Kennedy cut a high stump in 1942. They boarded up to avoid having a butt log with stump rot and root flare. Modern milling methods allow trees to be cut flush with the ground so no wood goes to waste. Photo from the author's collection.

PHOTO ILLUSTRATIONS

FOREWORD

There have been great changes in our forest lands since I worked my first day in a logging camp in February, 1923. We were clearing a railroad right-of-way and moving the logs with horses. We "sniped" —beveled—the top ends of the logs so it would be easier to move them. The foreman handed me an ax and pointed to the hillside covered with freshly cut logs.

Now, with the newly designed machinery, words like "sniping" are no longer used. When the steam donkey replaced the horse, new work and new words came into use. Wood-burning engines required men to cut the wood and fire the engines. When gas or diesel engine came into use, the men who operated these machines could control the fuel to fire the machine. These machines were all called "donkeys" and the men operating them were called "donkey punchers." The men who put the short cables called "chokers" around the logs were called the "choker setters".

The chokers were attached to the butt rigging on the main line. Each donkey had three drums on which the cables were wound—the main line, the haulback, and a small line called the "straw line". That last was pulled out into the "layout"—the area being logged. We then used the straw line to pull out the haulback line, which pulled the main line back to the logs.

In the 1920's the only way to move logs long distances was by water or by railroads. The development of new equipment cut down on the number of men that were needed. Many of the jobs, and the names for them, are no longer used in the woods. Electric signaling devices phased out the whistle punk's job. The steel portable tower yarder eliminated the work of topping and rigging up the spar pole. The heel boom, with its loading machine that moves on its own power eliminated the need for two men, the loader and second loader. When bulldozers and modern road-building and rubber tired log trucks arrived, out went the railroads and steam locomotives. The coming of the power saw made obsolete the hand saw and the chopping ax men.

Now, one man can get out as much work as three men could in 1923. The power saws were rapidly improved so that now we cut all the trees off down near the ground. We now cut all the trees. We no longer leave a tree that shows stump rot or any other kind of defect. These trees formerly helped to reseed the land for a new crop of trees. After a devastating fire, or the logging of an area, nature will in time reseed. But we get the next generation of trees quicker if we hand plant with seedling trees grown in tree nurseries.

Now we clear-cut our forest lands and take even the tree tops, down to two inches in diameter. Where the ground permits, we can go in and take out the windfalls, the dead, or the diseased trees, and leave our forest lands nearly park-like.

I well remember, working with the Ledgerwood yarders that reached out 3,000 feet. I remember Markham & Callow's logging in north Tillamook County with a railroad in the 1930's. The railroad went up the ridges, then reached out to near 4,000 feet each way with a skyline. When diesel-powered yarder engines replaced steam donkeys, fuel replaced the water that the old steam yarders had needed. These new yarder machines would yard up logs in two or three cold decks under this long skyline.

"Cold deck" is a term for logs piled up for later moving to the main landing on the railroad. Charlie Nettles, camp superintendent for Markham & Callow, called them "bunching machines." Loggers have their special languages to suit their work.

New machines are a wheeled tractor that backs up with a set of jaws to clamp on the log, and a carriage that runs on the skyline, radio operated by one man. It comes out to the log, bends down, clamps onto the log, and comes back to the landing. On signal, it drops the log. It doesn't need a choker man or chaser, that is, a man on the landing to unhook the log.

But on rough ground, we still have to get out the logs the hard way, be it with tractor, portable tower, balloon, or helicopter. When a new way is designed by one person, other men will soon develop better and more uses. Once we discover we can do the impossible, there seems no end to our accomplishments.

I will note when and where some of the new machines were developed. All through the Pacific Coast States, we have machines named either for their designers or for the places where they were first made.

The first steam donkey was named after the man who made it, but it was soon replaced by the Humbolt yarder. In Portland, the Willamette yarder was developed, in Seattle the Seattle yarder and the Washington log loaders. The Skagit steam loader originated in Skagit County, Washington. It seemed every small town had its factory for making logging or sawmill machinery. Here in Polk County, Oregon, where I live, the small town of Rickreal has a factory for making sawmill machinery. The tractor company at Independence, Oregon, designed the first blade to go on the front of a tractor. At Dallas the first lumber carrier and the forklift were designed.

Too often, large equipment companies have bought out the small companies and closed down their plants. The forklift replaced the lumber carrier. When a big company bought out the plant at Dallas and closed it down, jobs were lost, men laid off.

For years, we thought the vast jack pine forests in Eastern Oregon were only good for firewood. Then we found out jack pine could be made into chips. For a hundred years, the cutover forest lands in the southern and eastern states were growing young trees. When they were large enough to make small logs, a machine was designed to come up to these trees and cut them of at ground level. The same machine limbs them, barks them and chips them.

Now here in the West, our cutover lands are growing new trees large enough for small logs. Mills are built to process these small logs. The plant at Dallas, Oregon, now makes a machine to cut these small trees. A man-operated power saw is not needed on level ground.

The growing of new forest in the Northwest is called "tree-farming." Much care is used in selecting fast growing trees. Some strains of trees for making paper can be harvested after about twenty-five years. Firs grown for Christmas trees need less than ten years of growth. We can start cutting for lumber after about forty years for small saw logs. It takes about seventy years to grow trees suitable for plywood.

Trees grown on the gentle, lower slopes can be harvested by the machine that handles the whole tree. On the steeper ground, trees have to be cut into logs, then moved with the portable tower yarder.

Trees seem to grow best if we use seed from the higher elevations to grow trees to replant areas in those elevations. With careful management, our forest lands will forever grow trees to supply our needs.

Howard Brunson,
December 3, 1996

CHAPTER ONE

LOGGING IN THE 1920s AND 1930s
Written in January, 1977

This is how I came to take up the logging work.

In the fall of 1922, when I was 21 and considering marriage, my brother Royce decided that he would stay home and help my Dad with the farm work. My parents were willing that I should marry Olive Stanbrough and bring her home to live and then I could just keep right on working for them on the farm.

I had always planned to find a wife that I could live happily with for a long time. I had seen too many times when young married people tried to live in the same house with either of their parents; it usually led to problems that were hindrances to a satisfying married life. I knew that I would have to go find a job so we could try to make it on our own. In November of 1922 I bought a set of rings. I put the engagement ring on Olive's finger and then left the wedding ring in her care and started job hunting.

In those days all the big logging and construction companies hired most of their help through a private agency in a larger town. In Portland, most of the agencies were on Burnside Street. Some of the hiring offices were on Second Street. When they received the order for help, each office would write on a large blackboard the kind of jobs, where the work was located, and the name of the company. Men looking for work would walk the streets looking to see what work was available. We working men, when we saw a job listed on the board, went in and paid the required fee. We called it "buying a job on the slave market."

The men who managed these offices were more interested in the fee than our qualifications for the job. Sometimes there was a deal with the manager of the job and he would only let a man work a few days, then have a new crew of men sent out by the manager of the hiring office. Late November is a hard time of the year to find work. I went into Portland. For about a week, I tramped the streets looking for a job that I could ship out on.

Hurly-Mason Construction Company's logging camp in tents

I noted that every day one office had work listed for Hurly-Mason Construction Co. at Estacada on the Clackamas River. I found that Portland Electric Company was constructing a big electrical power project. They had 13 camps scattered along 35 miles of road with about 1,500 men working. Every day about 20 men were shipped out to work on this project.

One morning I got in line, paid my dollar fee, and signed on for common labor at 40 cents per hour. The regular streetcar took us to the end of the line near where the North Fork of the Clackamas runs into the main river. There we transferred to a truck. It fell my lot to be dropped off at Camp One just across the North Fork and between the truck road and the Clackamas River. I worked around camp at odd jobs.

The rivers were low and clear. The North Fork was about 20 feet wide with lots of large boulders. We made a foot crossing by laying planks across the smaller rocks. About a foot of snow lay on the ground when there came three days of hard rain. My, did we have a freshet in those rivers! The main river raised 18 feet and was full of driftwood that floated by at a speed of 20 miles per hour. Those big boulders, some 10 feet in diameter, came rolling down that river and trees over 200 feet long came floating down that 20-foot wide river, limbs, roots and all. It was quite a sight.

They sent me into the shop to help the welder. This made me good wages. We would get started welding on a big piece and sometimes worked 12 hours. There was no union so we worked at straight time and six days a week.

The camps shut down at Christmas time for eight days so I went home to Newberg. I found that train connections were such that it was not possible to get from camp to Newberg in one day.

When I went back to camp I was sent to Camp 1 1/2 and assigned to work with the pick-and-shovel gang. That freshet had taken out every bridge for 30 miles upriver from Camp 1. In some of the camps the buildings were of wood but most of them were tents. There were eight men living in each tent, sleeping in single bunk beds, two high. A wood stove for heat. Tarpaper roofing covered the board floors. Each tent had a canvas fly over the top. When the cold east wind blew down that canyon, all that canvas made a lot of noise. With the stove in the center of the tent there was no room for chairs. It made close living quarters for eight men. We were charged $1 a day for room and board. The food was good and was served in a tent that seated about 100. All the kitchen help was male.

The winter rains had caused many landslides. They cleared the larger slides with steam shovels. The smaller slides were all taken out by hand. Each man was given a long-handled shovel and when it was too far to pitch the dirt, he was given a wheelbarrow. I was sent out with one of these crews with a shovel. This may seem a slow way to move dirt, but you must remember that the big and powerful tractors with blades were at that time (1923) just an idea in someone's mind. Throughout the thirty-mile length of that road there were a thousand men with shovels and wheelbarrows attacking the slides that were blocking it. Every man who went broke in Portland shipped out on that job to use a shovel. Two young men had jumped ship in Portland, one from South Africa, the other from England. Three men came from Ireland, and one man said he had just spent $40,000 on a protracted binge.

One mountainside of about 30 acres of second-growth timber had slid out across the road and partially blocked the river. They had to move in a steam donkey and log off the trees before they could start clearing off the slide with a steam shovel.

Some of the married men who lived in Estacada could get into that town on Saturday evening and come back to camp on Sunday evening. These men all worked on the right-of-way clearing gang. They were paid fifty cents an hour. It seemed the logical thing for me to do was to get married and join this crew.

Well, things don't always work out the way we think they will. I laid off a few days and got married, then when I came back to camp I found that the right-of-way clearing crew was moving up to Camp 10. This was 35 miles upriver and at the end of the graveled road. We started working down the right-of-way near Camp 9 1/2. We went into this camp for dinner at noon. The cook put out a good dinner. We young loggers sure put away the food. We had a saying that we first ate to fill in the space left by the scanty breakfast that our cook at Camp 10 put on the table. Then we ate our regular dinner, and then we just ate to fill up for the food that our cook at Camp 10 put on the supper table that was not fit to eat.

Workers of the St. Paul & Tacoma Lumber Company at dinner in a cookhouse that seats at least 150 men. Photo courtesy of Tacoma Public Library, Richards collection.

I worked a week at brush-burning there. On Sunday a young second faller came down from Camp 11 saying that he had quit. I walked the three miles up the mud road to Camp 11 and got this job, which paid 75 cents per hour. Now that new pair of loggers' shoes was really paying off. Since buying my first pair of caulked boots my wages had jumped from 40 cents to 75 cents per hour.

Then again, things didn't work out the way that I thought they would. Here I was 38 miles up in the sticks with some of the poorest management that I ever had to work under. The camp push was following the orders sent out by someone sitting at a desk a hundred miles away.

I will explain about this right-of-way. The dam and head works were at Camp 12 and a pipe nine feet in diameter would carry water down to the powerhouse at Camp 9, a distance of about ten miles and a drop of about a thousand feet. At that time, the only way to transport those heavy sections of pipe was by railroad cars. They were making the roadbed for the railroad from the railhead at the North Fork of the Clackamas all the way to the dam site at Camp 12. The plans called for the railroad to be built parallel to the pipeline and on the uphill side. The orders were for clearing the railroad right-of-way first. This is a workable plan on nearly level ground, but not on steep hillsides.

On my first day at falling timber, I could see why they were not able to keep good timber men on the job. We were cutting through a thick stand of second-growth fir with a few old-growth dead trees with the bark still on them. The proper and safe way was to start at the bottom of the lower right-of-way, but the camp push insisted on the letter of the order. There we were falling those trees downhill and right into the standing timber and hitting those dead trees with all their bark still on.

Hystler locomotive in 1921 or 1922 at Black Rock, Willamette Valley Lumber Company. Engineer Art Kiel is the first man from the left. The other three men are unidentified.   Photo courtesy of Charles V. Ames.

My head faller was an older man and I was young and ambitious. I had visions of staying on with this large construction company and working up to foreman, or eventually the camp push job. In the afternoon of the fourth day we finished our section of the railroad right-of-way.

The next morning we started on the pipeline right-of-way. We had to use springboards to get above the downed trees. We no longer could stand on the ground. This made for high stumps. In the afternoon we came up to one of those dead snags. We put in our springboards, chopped the undercut, and were ready to start sawing on the back. I had that eight-foot falling saw and was waiting for my head faller to take his place. He had just used his axe to sight in the undercut to make sure that it was pointed right.

He was standing in front of the tree with his axe in his hand. I told him that I did not like the loose bark. It didn't look safe to me. I took ahold of the bark and gave it a little shake to test. Down it came in great strips. The last I saw of my falling partner was when a large chunk of bark fell straight down and tipped out. It fell on his head and drove him to the ground. With a whooshing sound, in a cloud of dust, all the bark came off that dead tree.

As I fell to the ground I managed to get my head behind a small tree that saved me from a bad beating. As I lay there with my legs pinned down, I lost a lot of ambition in the space of a few seconds. I had visions of my wife of less than a month getting a notice that she was a widow.

I managed to get free and started looking for my partner. The bark began moving and up came his head, all bruised and bloody. As he stood up, the bark falling from his shoulders piled waist high. I was so relieved that I laughed. All the crew who were working in sight or hearing came running to see how badly we were hurt. We were just plain lucky. We were both scratched and bruised but were able after a short breathing spell to finish sawing and falling the snag.

That evening after supper I told my head faller that I was looking for a safer place to work. He wanted me to stay, saying that with more experience I would be a good timber faller.

The way it looked to me was if I stayed on this timber falling job I had little chance of living to enjoy the companionship of a wife. When morning came, I put my few belongings into my packsack and walked the three miles down to Camp 10 and caught a ride down to Camp 9 where I was able to get my paycheck. I was able to ride a truck down to Camp 5. I ate dinner there at 12:00 o'clock, but found that the railroad was now six miles upriver from the North Fork so I had to walk the railroad track down to the end of the electric car line that went into Portland.

It was evening and after dark when I got out to my Uncle Wesley Brunson's place at Maplewood on the west side of Portland, only to learn that my wife, Olive, was at her Uncle Murray Hunt's home at Garden Home. This meant that I had another three miles of railroad track to walk before I could meet with my wife.

We determined that I should find work where we could live together. This was about the middle of March and the only work I could find where we could live together was wood-cutting for a dairy farmer on Yaquina Bay, about eight miles from Toledo, Oregon. We would have to buy a tent. It was too early in the year for camping out, but we bought a small tent and a sheet metal stove.

We took the Red Electric train to Monmouth where Olive's sister, Mabel, and her husband, Lee Payton, were living. Mabel was going to school and Lee was teaching in a country school west of that town. We stayed overnight with them, then went on to Corvallis where we took the steam train that let us off at a flag station called Storrs.

The dairy farm was across the bay from the railroad. The farmer came over with his boat and ferried us across. We set up a tent and the job started off all right. The farmer furnished the tools. I first cut up some drift logs with an old Vaughn dragsaw. This was one of the first dragsaws built by the Vaughn Company. The motor number was #65. The weather was mild.

After the first week things didn't go so good. Now, I had cut a lot of cordwood while growing up on the farm. There was a good stand of second-growth fir on this dairy farm, but I found that I could only cut trees too small for saw logs into wood. This changed the deal from what I thought were the original terms. Then a three-day rain storm put a stop to our tent living.

We decided to go into Toledo and try for work in the big mill there. The farmer was reluctant to let us go. He had advanced our train fare and groceries, and would not pay us any thing more for the wood I had cut, since I had only cut half of the wood that our agreement called for. We were only too glad to get away from that place, and we either had to get the dairy farmer to take us to Toledo when he took the milk in each morning or swim for it.

We rented an upstairs room in the hotel in Toledo. We cooked our food on an oil stove and sat on the bed to eat it. Olive made beds and cleaned rooms to pay our rent. I went to work in the big mill at 40 cents an hour. I never liked handling lumber, so after working a week in the mill I got a chance to work for the contractor building a new set of dry kilns. This work paid 50 cents an hour and we worked nine hours a day.

I still had it in mind to be a logger so, after three weeks at Toledo, Olive went home for a visit and I went out to the camp that was cutting the logs for the Toledo mill. C. D. Johnson had taken over this mill, which was started by the Pacific Spruce Corporation during World War I. This was a government-backed project to get out spruce lumber to be used for the construction of aeroplanes.

The railroad south ran 21 miles down the coast from South Beach on Yaquina Bay. This corporation built it to get into the fine stand of spruce timber near Yachats.

Manary Logging Company, mid-1923, on the Oregon Coast

The camp operated by the Manary Logging Company was six miles south of Waldport, Oregon, and three miles from Yachats. It was hard to get to this camp by car. You could ferry across from Newport to South Beach then go at low tide on the beach to the ferry across the Alsea Bay at Waldport, and then on the beach to Camp One.

This was the nicest logging camp at which I ever worked, all nice and clean new buildings. The cook house could seat 175 men and there were about 50 family houses. The smaller bunkhouses were designed for sleeping four men in single cots. The larger bunkhouses slept 12 men. The meals cost 40 cents or $1.20 per day. There was no charge for the bunkhouses, but the men had to pay for the bedding.

I took the steam train from Toledo to Yaquina and then the Newport ferry that stopped at the South Beach dock. It was a mile walk over to the logging company railroad. I waited there with several others and several packages of supplies until the log train came back with a flat car and a string of log trucks.

Each morning the big locomotive called Black Jack left Camp One with its string of 25 loads of logs for the dump at South Beach. It would be about eight o'clock in the evening by the time the train made the round trip to the log dump and back to camp. I got on the flatcar with several other people and the truck load of supplies. When we got into camp they told me to go into the cookhouse and eat with the train crew. I was anxious to see about a job. They said they would find me a bed in one of the bunkhouses and to worry about the job in the morning.

When morning came I went into the dining hall with the crew when the breakfast bell rang. They told me to wait at the door until one of the waitresses assigned me a place at a table. I noted that four young women were waiting on tables. I also noted that the men were respectful to these young girls and their table manners were good. Most of the men looked tall and much larger than I.

After a good breakfast I went to the office. The superintendent commented that I looked small for working with their large trees and thought that I would do better on the rigging. They sent me to look up the bull buck as he did the hiring of the timber fallers. This man, after talking awhile, said that Steve Rush was bucking but wanted a partner so that he could get back to falling timber. The wages for bucking were $5.50 per day. Second fallers got $5.75, and head fallers $6.

I went back to my bunkhouse and soon a big, strong-looking young man came in. After talking awhile he said that, yes, he would give me a trial as his falling partner.

Steve Rush was a second-generation timber faller. His parents were members of that group of people from Finland that homesteaded at Taft on the Siletz Bay and up Schooner Creek. Steve was very patient and taught me a lot about falling big trees. Up to this time I had worked only in second-growth timber. My natural way of chopping was right-handed. We usually had to go up two springboards high. This meant that some of our springboard holes had to be chopped left-handed. He impressed it upon my mind that a good timber faller had to be able to chop both right and left.

The trees were mostly spruce with some cedar and hemlock. Any tree that was less than 18 inches in diameter we left standing. I had to learn to pull a 10-foot falling saw. Each set of fallers had four spring boards, two five-pound White brand axes with 42-inch handles, a 14-pound maul, and four long slim tapered steel wedges that weighted 10 pounds. each, two sets of steel wedge plates, two eight-foot falling saws and one 10-foot falling saw. This was so that one saw would always be in the filing shed.

We were falling timber on the first slope of the mountain. We always rested a few minutes after falling one of these large trees. It was a new experience for me. We could sit on the stump and look right out over the Pacific Ocean. We had to chop most of the undercuts. This is usually the case on steep ground. The lower side of a tree is too timber-bound to saw into with hand-saws. That is, the saw would bind in the cut.

Paul Warner displays hand tools used in bucking logs crowded together in a draw in 1942. Photo by the author.

At this time, the gas-powered chain saw was twenty-five years in the future. These White brand axes were the best tools for making the undercut in the 1920 years. This was the middle of April, 1923. We always put a bend in our axe handles and the blade was filed flat on alternating sides. We used one side when we made the down slice on the undercut, then turned the axe so that the other flat side of the axe was down. This made the bottom of the undercut flat and level. The bend in the axe handle kept our knuckles from striking the tree when chopping a large tree. The timber faller would get bruised knuckles when he chopped deep into the cut. It would be almost impossible to chop into a large tree without this bend.

Many times I have had people ask how long it took a set of hand fallers to cut a tree that was eight or nine feet through. Some of the timber fallers only tried to cut one tree a day if they had to use their 10-foot saws. Steve and I averaged three trees in two days. A tree nine or ten feet in diameter takes from three to four hours just to chop in the undercut. The power saw speeded up this work. I have cut a tree eight-foot tree in diameter in thirty minutes with a power saw. Cutting the felled trees into logs, called "bucking," took several times as long as it does today. Power first saws weighed as much as some of the loggers did. Now they are light enough for one man to operate.

The Whitney Logging Company

The Whitney Company had acquired timber holdings on the Kilchis River in Tillamook County by 1923. They built a sawmill and a town on Tillamook Bay at Garibaldi. They moved a lot of equipment from their operations in the deep South. They also brought along some of their older workmen. The large trees and rough mountain sides presented problems that their men and equipment were ill-prepared to solve.

They tried to log downhill. The most practical plan is to have the landing on the top of the hill and haul the logs up to the spar tree at the landing, and the yarder tree on the hill. At one place they built a dam in the river and logged a hillside to the pond. Then later they loaded the logs onto rail cars for transportation to the log dump at Idaville on the bay.

This dam was so constructed that the salmon could not get past it. A dam like that would not be permitted in later years. This was a good example of the little loggers of that day cared for the wildlife in their forest holdings.

In the late fall of 1923, I needed a winter job. I had spent my summer's wages on a car, and since my wife was pregnant it was time for me to find steady work.

I went into Portland, and the only job available was chokerman for the Whitney Company at their Kilchis River camp. When I got out to their main camp I found that two of us were to work at the incline camp. This incline went up a steep mountain. The road grade for the railroad track would be 3,300 feet long to the little flat where a small camp was built. This railroad climbed to an elevation of 1,500 feet above its starting place.

It was late afternoon when we arrived at the base camp. They showed us a trail to take to get up to the incline camp, telling us to hurry or we would be late for supper.

Charles Taylor resting after bucking, with hand tools, a log almost as thick through as he is tall. Photo by the author.

When we finally gained the upper camp, we found the men had a bunkhouse to sleep in but the dining room was just a tent lit with lamps. The flapping of the tent in a storm often blew them out while we were eating our meals.

They were using a steam yarder for the Bagley scraper that made the railroad bed. They were making the last cut at the top of the roadway for the rails. A section was to be double-tracked at the halfway point. The empty car coming up could thus pass the loaded car coming down. I soon found out why they were short on rigging men. It took a lot of wood to keep that yarder engine in wood fuel. They had wood split, but one man could not make the four-foot lengths of wood fast enough to keep up with the demand. It didn't require much time to change rigging and we choker men were expected to help out with the wood splitting. The prideful other choker man thought it beneath his dignity to split wood. He soon took off down the mountain.

Well, I had cut lots of wood on the farm, and I came to work. In due time we completed the grade. There was a crew laying the ties and rails, working their way up the mountain. They used another yarder to haul up the ties, kegs of nails, and rails on a specially built car pulled by the other machines. They had one main line that went out to a side block, then up the mountain to a lead block, then the line came down the right of way to be hooked to the materials car, to pull it up to the top of the incline. This incline was quite steep.

About midway, where the passing track would be, there was what we called "the flats". Here we took our main line straight down from the upper machine. Two of us would take the main line and start down the mountain. The donkey operator would have to use the brake to hold the line at our pace as we ran down the grade to where we could wait for the material car.

In those days the crew used a whistle wire for signaling the donkey operator. This wire attached to the whistle we took down to our work place. Since I was the youngest, it fell my lot to work the whistle wire. We made the end fast. Then with a sharp jerk, if the wire was just right, the whistle on the donkey would whistle out the signal and the donkey operator would manipulate the lines as desired.

Just a day before we were to shut down for two weeks at Christmas time, we had trouble. The old fellow who operated the lower yarder donkey kept complaining that the friction blocks on his machine were worn and ought to be replaced. The foreman thought he should make them last one more day. We at the upper work site knew that the donkey operator was having trouble getting the loaded material car up to where we could hook onto it. We could see the spurt of steam from his exhaust as he gave it a final try. Then the loaded material car started going back. There was a layer of fog near the river. Soon after the car dashed out of sight in the fog, we heard a loud noise. Our foreman went down to see what damage was done. When the main line came out of the drum of the yarder, it caught in the side block and stopped the material car. It kind of broadcast the rails and the tie nail kegs. They broke open.

Fortunately, all the men were able to get to safety. Old Herman, the donkey operator, took one look at a tie that had jammed into where he had stood until his foreman had literally jerked him in back of the machine. Old Herman grabbed his lunch pail and started walking, saying, "I am needed in town." He also called his son who was working on the crew. He said, "Come on, son, let us go see your mother."

Since we could do no more work on the incline until repairs were made, we all were told to start the Christmas vacation.

I never got to help finish this project. In due time the track laying was finished. A specially built snubbing machine was hauled up the mountain. A small locomotive was brought up. Logs were brought down, and as the car loaded with logs was let down, the empty came up. They passed at the halfway point on the passing track.

The Whitney Company had only logged one landing at the top of the incline when the management decided to abandon the Kilchis River country and get their logs from the Nehalem River country. The only trouble was that the equipment to log this rough country had not yet been designed. Twenty-five years later another company with modern road building machinery built a road up into and logged the timber that this incline had been expected to reach.

When we came back to work after the Christmas shut-down, the hook tender, as my foreman was called, brought two experienced chokermen with him. He said he thought I would get along better on the cutting crew. I only had to see the bullbuck to get work at the main camp.

One surprise met all the crew when we came back to work. A notice at the office stated that all our wages were cut twenty-five cents a day. We had two choices. We could go to work, or we could go back to town. I went to work with a bucking saw for $5.50 for an eight hour day.

We often laugh at a near accident, not because it was funny, but in relief when we realize how serious it could have been.

There was the time that Mike Hoffman dodged a boulder. Mike was a large, good-natured Irishman. The yarder was up on the top of the hill. The logs were first pulled up to this spar tree, then brought down on a skyline to the landing tree on the railroad. This line brought the logs down a draw where, when it rained hard like it can do in our Coast Mountains, a small stream poured.

Mike was up at the landing when the quitting whistle blew. The crew all grabbed their lunch pails, or "nosebags" as we loggers called them. The quickest way down to the landing where the crummy waited was down the draw and its mud and water. Mike took his time. As boss it was necessary to show some dignity before his crew. All the men were on the crummy waiting for Mike. While he was still some distance from the lower landing, a good-sized boulder, loosened by the rain and running water, came bounding down the mountain. Some of the crew men started yelling to warn Mike. When he finally sensed his danger, he started running and looking back. It looked like the boulder would strike Mike. Then he stumbled and fell head first, sliding in the muck. The builder bounced over him and rolled harmlessly against the logs at the landing. Mike came on down to join the crew on the crummy. He greeted his men with a smile, saying, "Boys, if I had not thought quick and laid down it would have got me for sure."

One of the misfortunes of the Whitney Company was losing the yarder off this hill. In yarding with a spar pole after working half of the circle, the high lead block must be switched to the other side of the spar pole. The yarder machine must also be moved to the other side of the spar pole. It seems that always an accident is just waiting. One wrong move and it happens. There was a stump that should have been blown out of the way of the yarder, or a safety line put in place to hold the yarder, since the hill dropped off sharply. To get around the stump the heavy machine would go dangerously near the drop-off, and over a rock cliff. The hook tender, in his haste to get back to logging, failed to take these safety measures. When the machine started slipping toward the cliff, there was no safety line to stop it. The crew just had to get in the clear and watch their machine slide over the cliff and crash to pieces. The drop-off to the bottom of the canyon was more than a thousand feet straight down.

*American saddle tank steam locomotive, fired by oil.
It's water tank rides atop the boiler instead of behind
it. Photographed in 1934. The man is George Graves.
Photo courtesy of Charles V. Ames.*

The 1923 I.W.W. Strike

With working conditions and pay so much improved by our strong labor unions and binding contract agreements that both the company and the workers respect, it is hard now for all us to realize the unstable working conditions of the 1920s.

I was working for the Manary Logging Company at their headquarters camp twenty-one miles south of Yaquina Bay and three miles from the present city of Yachats, on the Oregon Coast when the I.W.W. was trying to organize some of the logging crews. I had only worked three weeks as second faller to Steve Rush when they called a strike.

At that time you had no job security. A foreman could discharge a workman for any or no cause. In order to get work for the larger logging or construction operations, usually the workmen went through an office in town. Portland had a number of these places near Burnside Street. Most of the idle men spent their time there while in town. When a company needed men, they posted job descriptions, with the wage to be paid, on a bulletin board at the Portland hiring hall. We workmen paid a fee and usually were advanced the fare out to where we could get onto the company's private railroad.

Often two men who were friends would ship out on a job. If the foreman for any reason disliked one man, he would also fire the other. Some foremen used the phrase, "Take your buddy with you." On a great many jobs, the company furnished the bunk house, but they expected the men to bring their own bedding or buy it from the company commissary.

When the Union called that May Day strike, some of the men did not want to be named "strikers." They left camp a few days before the first of May. My falling partner, Steve Rush, decided he was overdue for a visit to his parents' home. Now it was up to me to learn how to cut those large trees into logs. I had no wish to stop work. They gave me a set of bucking tools and I kept on working. On the morning of the strike about a quarter of the 175 men who ate at the cookhouse called for their time. They got ready to ride the flat car hooked just back of the locomotive that pulled the first train of logs to the log dump at South Beach, Yaquina Bay. There they could get transportation to town.

I had no reason to leave this job, so I went to work with the others who chose to stay on the job.

It was hard work learning how to cut those large trees into log lengths. I worked until September, then when prune drying time came I decided to find other work. Though the food at the cookhouse was good, I was worked lean and I felt like I needed a rest. I went home to mother's cooking and lots of fresh fruit for the picking. I gained nine pounds in weight the first week.

One thing the strike gained was that the company furnished us the bedding and hired maids to keep the bunkhouses clean, and to make and change sheets on our beds when needed.

Coats Logging Company

The Coats Company had their mill on a slough at the north edge of the city of Tillamook. Their logging holdings and camp were about six miles south of town. I first worked for this company in 1924. I worked for them several times in later years. This company had the reputation of paying low wages. With the beginning of the depression in 1929 they, like many other logging companies, just left their equipment out in the woods with little or no care. Much of their machinery was old and badly in need of repairs, or later sold for scrap metal.

The Coats company logged from a railroad about eight miles long. When times looked better in 1933, they bought a tractor and started hauling their logs direct to the mill with trucks. They had one almost-new yarder when they shut down their logging. They decided to yard a unit with this yarder on the railroad.

In late August of 1933, word went out that they would need a cutting crew. Claude Evans would have charge of the falling and bucking.

When I, with a partner, went to see Claude Evans about the work, he said the price would be 24 cents a thousand for the felling and the bucker would get 20 cents a thousand board feet. The pay seemed low, but we knew that the minimum wage had been set at $3.20 an eight-hour day. There was no other work available. We decided to give it a try. My falling partner, after two weeks, decided he would go to Hood River with his wife and work in the apple harvest. We found that at 24 cents a thousand feet for falling trees, our own pay came to only $2.80 for an eight-hour day. They had paid him off at the rate of $3.20. We were having to drive our car the 30 miles to work at our expense, besides feeding our families. I worked another week, then was offered work close to home at that same wage, so I drew my pay and quit working for the Coats Company.

The other members of the cutting crew lived closer to the work, so most of them stayed until they could get a payday. There was no draw day, and most of the workers had to work until October 15 to the one payday each month. Most of the men were paid at $3.20 a day. Three men on the crew earned nearly $4 a day. Most of these men had bills and some bad debts of long standing. When payday came and these men went to get their checks, they found that the collection agency in Tillamook had their paychecks. Since the men still had pay for two weeks work coming, the only way they could have money for their own was work two more weeks, then quit and draw their pay. The unit for logging was cut and the Company no longer needed 20 men on the cutting crew.

After times got better the Coats Company had to pay better wages. By 1936 their wage scale was up to the rate set by or negotiated by our union. Claude Evans was good at this work and knew how to get the timber cut. He often had to use young and inexperienced men. He could always use an experienced man. In later years I worked for him several times. When I needed work, I would go talk with him and tell him that I needed work, and he would usually say, "Come to work day after tomorrow. I have a young fellow who should not try at this kind of work and you can use his tools."

Claude Evans appreciated a good timber faller. I remember one time I was cutting trees on Jordan Creek, way up on a side hill above the truck road. Some of the company officials came out to check up on how the work was progressing. This group of men just happened to come by as we were getting ready to drop a large, leaning tree. This tree also had a side lean. As the men came by, Claude stopped them at a safe place where they could see this large tree fall. Claude pointed out to them the spot where the tree should hit the ground he called the "lay." He pointed out how the tree leaned to the right side of the lay and how the back cut would be made so that the tree would swing left. They watched as the tree slowly moved until it was in line with the lay. They heard our cry of "Timber!" Claude said, "Now he will cut it loose."

The men watched as the tree fell nicely into the place where we had willed it to lay. The tree scaled out at 19,000 board feet. One never knows who is watching us as we work.

Moving the yarder at Coats, and the locomotive that crossed a falling bridge.

When we went to work on Labor Day of 1933 for the Coats Logging Co., the unit we cut was about five miles out on their eight-mile railroad. They still had a small locomotive and a boxcar with benches to haul the crew to work. Just beyond the landing was the high railroad trestle across the Tillamook River. The yarder they were to use was out another three miles and across another bridge. They would let the cutting crew off right at the first bridge. Then the crew that was getting the yarder ready to move went on out to the old landing.

After we had worked a week, this crew said they hoped to get the yarder ready to move that day. Soon after noon we heard a terrible noise and wondered what could make that kind of noise. That evening when quitting time came, we of the cutting crew stood by the railroad track and watched for the little locomotive with the crummy to come to take us down to camp. Right on time, here came the crummy. At the far side of the bridge it stopped. The crew came walking across the bridge. Then here came the little locomotive creeping across the bridge with no one in the cab. After it had cleared the bridge, one of the crew climbed in and stopped the engine. Then here came the engineer walking across the bridge. We soon learned that the other bridge had collapsed under the weight of the yarder and the special, strongly made flat car that carried it.

The way that it happened was this: When moving a heavy load like this yarder machine, a flat car is put between the heavily loaded car and the engine. The approach to this old bridge that had stood several years without care was a turn with a low bank on one side and a canyon on the other. The side next to the bank gave way first and the yarder slid off unharmed. There was that little locomotive trying to cross and the bridge falling behind it. It just sat there a moment with its drive wheels spinning until the tilting of the cars broke the coupling. Then little engine scooted across to safety.

It took three weeks to move that yarder out to where they could get it to the new landing. If that bridge could have stood up to the load, it would have been only a thirty-minute move. Never again would the crew ride across that remaining railroad bridge. They always walked across, ahead of the locomotive.

Shevlin-Hixon Lumber Company
    and logging in the pine timber

In 1925 I worked in the pine timber for Shevlin-Hixon Lumber Company out of Bend, Oregon. Today they have one of the two largest lumber mills in the United States. Even then, their mill took 40 acres of timber a day to keep it going. They were still using some horses to haul out logs. They put the front end of a log upon a carriage with high wheels to keep the log from scooping up dirt as the horses pulled it over the ground. This made it easier for the horses to pull. This logging in the pine forest was all on level ground.

In 1926, soon after that New Year, they replaced the horses with tractors. The tractors had metal arches that lifted the front ends of the logs up from the ground. Shevlin-Hixon's main logging machine was a Ledgerwood yarder and loader. This machine sat astride a railroad track, high enough for the empty rail cars to come right under it to be loaded. It was tower-like. Cable ran from the top of this machine out to the top of what they called the tail-tree. A car ran on this cable, so that the logs were picked up and carried through the air to the loader. This machine had two motors, two sets of drums. One motor ran the yarder that hauled in the logs. The other ran the loader to put the logs onto the railroad cars.

The author ringing the quitting time bell during World War II. Watches were unavailable. Not every logger had one.

Photo from the author's collection

Flora Logging Co. and the Tillamook Burn fires

When I first started working in the logging camps, most of the logs were hauled on railroads, and the only way to get out to the camps was by the logging company's own railroad. I soon found it paid best to work for the larger logging companies. At times I have worked for the small outfits called "gypo loggers", or small mill operations that were usually under-financed. Sometimes you waited for your pay, and sometimes they never paid. In those days, there were few laws governing wages and insurance compensation for injuries. We worked six days a week with no paid vacations, no overtime, and no unemployment insurance.

When I worked for the Flora Logging Company, west of Carlton, Oregon, it took six hours to go from Carlton up to the camp. Usually, Sunday was a rest day, but there was always work for those who cared to work. Some of the men regularly worked every day for three weeks, and then they would take four days off to go to town and visit their families.

Flora Logging Company had three camps, two with cook houses that fed more than 100 workers each. The headquarters camp, a smaller camp, had mostly family houses for the railroad crew and the supervisory personnel. For use around the camps, the company issued money tokens. The bank at Carlton would exchange this company money. All three of these logging camps were burned in the 1933 Tillamook Burn fire. Everything was lost. The crews managed to escape with only the clothes on their backs. All of the railroad bridges, which were made of wood, and all of the logging equipment that had wooden parts, were burned. During World War II, when scrap metal was in demand, bulldozers were used to make temporary roads into the Burn. The steel rails and steel from the logging machines were salvaged. They found Flora's fully rigged spar tree still standing with the logging cables out just as the crews had left them. Most of the fir trees were cut in later years, due to the demand for plywood and peeler logs.

When falling trees in the Tillamook Burn, we worked at piece work. But when they started taking off five inches for rot in the diameter of the fire-killed logs, I started looking for green trees and went to work in the Cascade Mountains.

It seemed I could always make more take-home pay by working at logging. I tried farming, but it never paid off. Either the berry crop was poor, or the price was down due to the Depression in 1929 and the early thirties. The farm was owned by Olive's parents. It was on Chehalem Mountain, about two miles northeast of Springbrook, Oregon. We lived on this farm for three years. In those days the doctor went to the patient. Paul, our third child, was born in the small house we built on that farm.

My mother was a good midwife, so our first two children, Marjorie and Virgil, were born in the house on my parents' farm at Mountain Top, on Chehalem Mountain.

Aftermath of the Tillamook Burn of 1939. Forest fires destroyed tens of thousands of acres of Oregon Timber in the 1930s and again in the 1940's. Many of these burned tree trunks have solid wood in their centers, worth salvaging.   *Photo by the author.*

We moved to Rockaway, Oregon, in October, 1930, and lived there until 1943. Then we moved to Siletz, Oregon, for two years, then back to Rockaway for four years. The family stayed at our house at Rockaway while I worked for a year out of Eugene, Oregon.

The Depression started in the fall of 1929, and work was hard to find for the next four years. I worked for O. B. Randall whenever he had work. He rebuilt small bridges for Tillamook County and ran the county rock crusher at Roy Creek, at the head of tidewater, on the Nehalem River. My wages were two dollars a day.

By 1935, the logging business had picked up considerably. Wages increased, we were able to find steady work, and we were able to pay our way. The falling and cutting into logs of large, old-growth trees requires skills that take years to learn. Several winters, when the logging camps were shut down, I cut cedar light poles for the electric power company. The late thirties and early forties were a good time for an ambitious workman to get started on his own. The price for stumpage was low.

Those burned trees could be bought for one dollar per thousand board feet. A good way to get started was for two men to go together. They would contract to buy a small tract of trees, fall and buck the trees, then hire a tractor or donkey engine to load the logs on trucks for hauling to the mill. Several men I worked with made it good. Some didn't make it so good. I was one of them who didn't make it at gypo logging.

Horses like these replaced oxen for yarding out logs in the larger logging camps early in this century. However, a few small logging operations still used horses until well into the second half of the century. Horses were more intelligent, more tractable, and cheaper to feed than oxen. Here, a team of four horses pulls a turn of logs on a corduroy road.
Photo courtesy of Washington State Historical Society.

Horse Sense

I learned about horses in 1929 while getting out cedar poles on Chehalem Mountain in Yamhill County. It is surprising the way horses show their intelligence. Harold Lamb yarded out cedar poles with a pair of small horses. These horses would obey his spoken word. Sometimes to get a pole to move, both of us would have to help with our peaveys. When we needed the horses to be pulling on the pole, Harold would tell them, and they would pull with all their might until he called them to stop. I have watched these two animals pulling when they were attached to a cable 100 feet from where we two men were straining to start the pole from the object it was lodged against.

In 1933, I was yarding out cedar poles for the electric company in Tillamook County. A man by the name of Howard, a direct descendant of Sam Howard, one of the first settlers in the Tillamook Bay area, was yarding out alder pole logs with an old black horse. Our practice was to first take out the cedar, then the alder, and then fell the old-growth fir trees.

Heel boom loader lifting a log with a center hook at Jones Logging Co. near Kernville, on the Oregon coast about 1948. This type of crane is still used in today's logging. The machine is mounted on tracks for easy moving to new sites.   Photo from the author's collection.

I hired Howard with his horse to yard out the cedar poles to my landing on the county road. I was surprised to find out that his horse needed only to be led on one trip from where the pole was cut to the landing. Howard would get the pole ready and hitch the horse to the pole. If the pole hung up, old Jack would pull first to the right, then to the left until the pole came free. I stayed on the landing. When the horse came to the landing, he would stop right where I wanted the pole. I unhooked him from the pole and he would go back to where his master had another pole ready.

Howard and his horse were still yarding out the alder when a crew came to cut and log the old-growth trees. It was a sad day when someone missed the intended place where they wished the tree to fall. The top of that tree hit the faithful old horse and killed him. Howard seemed to feel as bad as if one of his sons had been killed. We grieved with him, and I regretted that I could no longer have the use of this intelligent animal.

Note: As recently as July, 1977, Ray Hanson of Vashon Island, WA., logged alder with his gentle, white giant of a horse, a Clydesdale named Barney, who would carry as many children as his back held. MW

The Jones Logging Company, 1950

I did falling and bucking by contract. I was usually paid at the final settlement on the scale of the logs when trucked to market. My longest cutting job was for the Jones Logging Company at Lincoln City. I had only a verbal agreement. I started in November of 1950. They had only one yarder to begin with. They kept growing until they had three yarders and two tractors. I didn't get a final settlement until almost two years after the termination of this agreement. At one time we had twelve men on the cutting crew. I furnished all the tools for the fallers and buckers.

Murphy Logging Company, Cottage Grove

The largest cutting contract I had was with the Murphy Logging Company at Cottage Grove, Oregon. At one time we had 22 men on the cutting crew. The men furnished their own tools and were paid by the piecework. I was supposed to get the final settlement on the truck scale, but there were so much rot and so many low-grade logs that I was paid $600 a month plus expenses. I found that this company was always good and fair, if the workmen did good work.

At the termination of my contract with the Murphy Company, I tried gypo logging but it didn't pay off. Then the state welfare people put a lien on our property, to collect for the welfare payments made to my parents, who had lost their farm in the Depression of the 1930s.

The only thing I could do was close out my logging business, sell everything we owned, and start out with only $1,000 left after settlement.

Nelson Logging Company

In the summer of 1925, I worked for Frank Nelson on Baker Creek, west of McMinnville, Oregon. He built a chute about 2,000 feet long to bring logs down a mountainside. They yarded the logs into the upper end of this chute by a steam donkey. Until the mill pond was completed, the logs landed on the ground. Sometimes the logs shot down so fast that they didn't stay in the chute all the way down to the water.

In the railroad logging days, if a mountain was too steep for the Shay locomotive to climb, they built an incline. The loaded log cars were let down the mountain by a cable from the yarder machine. Gypo loggers thought any steep road they could get an empty log truck up was not too steep to send a loaded truck down. Such roads were hard on truck tires.

I remember, in 1942, a road up to the landing was so steep that to get the panel truck up we had to have the weight of six men or three barrels of gasoline. If you had nothing else, you piled rocks into the back of the vehicle until you had enough weight to provide the necessary traction to the rear wheels. We always called that steep road the "Bart Road" after the man who first took a log truck up it. Later an easier, but much longer road was built up the mountain. World war II brought a stop to the hauling of logs down those steep mountain roads. When truck tires were rationed, the loggers had to build more roads with less steep grades.

Pope & Talbot Co.

We moved to Oakridge Oregon, in 1962, and I went to work cutting trees for the Pope & Talbot Company. At last I was free from all business problems. We really began to enjoy life. When I got off the crummy at the end of the day, I was free to work in our garden or cut and polish agates. On the weekend, with no loose ends of business to catch up, I could go agate hunting.

One change in working conditions was that when the logging changed from railroading to trucking the men first had to furnish their own transportation. This led to car pooling. When World War II started, with gas shortages, the companies had to start furnishing transportation. Most main trucking roads were paved. The men lived in town and the company ran buses from their homes right out to the landing, often 60 miles from their homes. Now almost all loggers commute to work.

The development of modern logging equipment has drastically changed methods of getting logs to market. When Pope & Talbot logged in the Oakridge area, you could all kinds of newer logging equipment in use in the Cascade Range. You could see all types of tractors, telescoping towers, and on the

*The "Climax" engine, first built as a wood-burning locomotive but later converted to
burn diesel fuel, shown as a wood-burner in 1922 on a switchback railroad with a
trestle bridge in background. Photo courtesy of Washington State Historical Society*

steeper slopes, balloons and helicopters at work. No longer do the logs go to the mills by rail; log trucks
have taken over. Highway 58 between Oakridge and Eugene was more heavily traveled by log trucks than
any other highway in Oregon. On the 44 mile-mile stretch of road from Eugene to Oakridge, we normally
met one loaded log truck for each mile traveled.

During our time at Oakridge, my late wife, Olive Brunson, developed her talent for making porcelain
dolls, selling doll kits by mail. Her dolls went to every state in the Union and to several foreign countries.
I became a rockhound and with the late George Willey, another retired logger, invented the Willey B
rockhammer to mine agates. I made jewelry, bookends, and other agate objects.

I worked on cutting crews at Pope and Talbot until my retirement in 1967. I was a watchman for
C & K Logging Co. for a more few years, then watchman at West Fir Mill until it burned down in 1979
while m;y wife and I were on a short vacation. I always felt that had I been on the job that night, the mill
need not have burned down.

In the next two chapters I will tell about many of the loggers I knew during my almost half-century of
working in our forest lands.

CHAPTER TWO

ABOUT COUGAR BILL AND OTHER TIMBER BEASTS

When logging was by railroads built into the timber holdings of the large companies, the crews usually stayed in camp months at a time. It was not work that a man with a family would want for steady employment. Most of the crews were single men. Men working and living this kind of life, when in town and spending their hard-earned money, gave the loggers the reputation of being wild, rough, and tough.

Not so for all of these men. There were always a few men with families. Some of them would work every day for three weeks or a month, then take a few days to go home and see their families. I remember that in March of 1924 I worked 27 days at the Whitney Camp on the Kilchis River before I took time to go and see my wife at Newberg, Oregon, where she was caring for her stepmother while she recovered from an operation.

With the making of roads for hauling logs out of the woods by truck, this all changed. With a normal work week of five days, a man could get home on weekends. If he so chose, especially in the season of longer daylight time and in good weather, he could drive his car to work and be at home every night. In the winter, and the stormy time of year, I often chose to stay at the camp and only come home for the weekend.

I will describe some of the men I worked with in the logging camps over the many years I worked in the Oregon forests.

Cougar Bill

It was in 1932 that I first met a man called Cougar Bill. He lived in a small cabin on the bank of the Nehalem River a short distance upstream from Cook Creek.

I found an unusual way to cross the river a short half-mile above Lost Creek. A fir tree had blown down at a place where the main flow of the river cut against a bluff. In sliding down the steep mountainside, most of the limbs had broken off. The bark was still on the tree. We found that we could wade[1] out to this tree and, using the limbs as handholds, we could climb up and cross the river when the water was at a low flow. It was a rather tricky task to climb down this ladder tree with a pack of deer meat in hunting season.

I knew that Cougar Bill had a boat at his cabin. I learned that if he was at home he would ferry hunters and fishermen across the river. I had made only two trips on this tree crossing to find that I best should meet and start a friendship with this man.

Cougar Bill lived a primitive lifestyle. He trapped fur-bearing animals and hunted cougar. There

---

[1] Note: In wading a stream, we could go in deeper than our boot tops as the swift current will hold the pant leg tight against the leg. By tying a string below the boot top, with our rain pants we could wade crotch-deep in water and not get water in our boots. It also helped to use a walking stick.

*Preparing to load a 7000 board foot log using a tractor crane at Jones Logging
Company near Siletz, Oregon, 1953.   Photo from the author's collection.*

was a bounty paid for cougar at that time. Cougar Bill mostly lived off that wild country. He needed
only a few staples articles of food. Only two families lived at Batterson at that time. He had no car,
so he either walked the six miles to the Mohler store or trusted to his neighbors. It was nine miles
from Bill's cabin to Wheeler.

Cougar Bill had two hound dogs. These dogs were trained to track cougar, bobcat, coyote, and
bear. Bill also had a small flock of sheep. I noted that he had no shelter for his sheep. I asked him
about this, and he said that he hoped to develop a strain of sheep that could survive without shelter
from the winter storms. I doubted the chance of their survival when I noted that many of them had
lost all the wool off great patches of their backs. On a frosty morning we would see white frost on
their backs.

Bill, like his sheep, never cut his hair or shaved in all the years that I had his acquaintance. It
made him look much older than his actual age.

Late in 1935, I heard that the Hammond logging camp planned to do some logging. I drove up
to Foss. It was on a Friday, and I learned that they planned on starting a small cutting crew on
Monday. I also found out there was a new deal on for workmen. When I signed on for the payroll, I
also had to fill out an application for a Social Security number.

I decided to also go and see Cougar Bill and to go up Lost Creek and catch a limit of trout. I
drove up to Batterson and when I called, Bill came and ferried me across the river.

Bill was quite a non-talker, so after a few words I went up to Lost Creek. I fished up to where
the creek had been covered over at some far distant time by a vast rock slide. It had covered the
stream for about a quarter mile. The stream somehow made its way underground. This is why the
creek was named Lost Creek. Above this vast slide, the creek went another two miles and was full of
mountain trout. Near the slide I found the remains of a sheep probably killed by a cougar. When I

came back to Bill's cabin I told him about this varmint-killed sheep. I expect this was the main
purpose of the sheep—to attract the predatory varmints.

Bill started planning that next morning he would take his dogs and go check on his killed sheep.
I had never hunted nor even seen a live cougar. I had often hinted to Bill about taking me on a hunt.
Now I had him where he could hardly say no. He agreed that I should be there at daylight. We would
travel light, he with only a pistol. He never carried a lunch. When he killed a varmint, he would build
a fire and he and his dogs feast on the meat. He claimed any of these animals would be good eating,
with cougar and bear his first choice.

When I left home that morning, I told my wife I didn't know how long the hunt would take. Bill
thought that if it was a cougar it would go up to a nearby rock cliff and lay until it needed more
food.

I was ready to cross the river before daylight. Bill was watching for my car lights. He was ready
with the boat by the time I got to the river. I did put two candy bars in my pocket and brought a
flashlight. I had learned to always carry a flashlight on a hunting trip. If caught in the dark, or if I
needed to build a fire in the rain, I was prepared.

We went the two miles up Lost Creek to the kill, then took off up the mountain-side to the ridge
top. The dogs were cold-trailing an animal, but it didn't seem to be the one that had killed the sheep
on the creek. They searched out every rock bluff like they expected to find a deer lying in the shelter
of an overhanging cliff. When we got up to an elevation of 2,000 feet we were into a stand of green
trees. The 1933 Tillamook Burn fire left lots of green trees at the higher elevations. There were
patches of early snow, the wind was quite strong on the ridge tops, and some rain fell all morning.
On our Oregon Coast, near the time of the low tide, a mild rain and windstorm will often stop, and
out comes the sun. Then it may cut loose and blow so hard you had best start for home or seek a
safe shelter. I knew that by noon we could expect a change and, sure enough, it was for the worse.

That cougar had led us up to near the base of Cedar Butte. All the ridges head up to Cedar
Butte. The streams that start at the base of this mountain curve in more than a half circle. Now in a
storm, you could easily take a ridge that would take you in the opposite direction from your camp.
Cougar Bill seemed to know where we needed to go. He had hunted all over this mountain. When
the drenching rain started, with a wind that was blowing down trees and filling the air, he found a
trail that led us down into a canyon that I trusted would take us to the Nehalem River. When we got
down into the burned timber, the bark was flying off those fire-killed trees. Limbs and top sections
and whole trees were falling all around us. I could only follow Bill and hope to keep in sight of him
and his two dogs.

It was mid-afternoon when we finally came down to a good-sized meadow. Here was a rough
cabin. I expect that this was the place he headed for when the wind drove us off the mountain. We
sure were glad to get in out of the rain and wind. We were dressed for winter with wool underwear
and socks, loggers' boots and rain-proofed heavy canvas coats and pants. We called them tin clothes.
When the waist-length pants were put on the floor belt side down, the legs stood up straight.

This cabin was only 12x16 feet. The roof and sides were made of 36-inch long split shakes. The
floor was split cedar in six-foot lengths about three inches thick and two to three feet wide. A
window on one side was boarded over with cedar shakes. Under the window was a rough bench-like
table of the same material as the floor; on the opposite side, a crude bunk bed of split cedar. There
was no stove and no chance to build a fire to warm the place. However, it was dry and a safe haven
from the wind. Bill said we would have to stay there until the wind eased off. We could only safely
go to the river. It was less than a mile up over the mountain, but with the steep hills along the river,
past where the ladder tree had been. It would soon be dark. We knew we had best just tough it out
until morning. I was glad to have put those two candy bars in my pocket.

When the Balderee Logging Company took over the logging at Foss, Oregon, it was a one-landing camp. There were only 15 or 20 men staying in camp. I was staying in one of the bunk houses. Just a few days after our cougar hunt, Cougar Bill stopped to see me one evening. He told me he had gone up toward Cedar Butte and got that cougar. I noted that he had his pack full of groceries. I asked him, "How did you get to town to get your groceries?"

He replied, "I have an arrangement with the cook."

I had wondered about some of the meat we were served. I knew that we ate meat that had not come from a butcher shop.

I will never forget the shock and dismay we felt when the news came over our radios of the destruction caused by the Japanese attack on Pearl Harbor in 1941. None of us ever expected the change it would bring to our lives. In a few short months most of the younger men were in the armed services or war-related industry. We people living on the Oregon Coast were required to completely black out our homes. We could not let a crack of light show. Our streetlights were turned off and the store windows closed tight at night. When we could get gas enough to travel at night, we had to creep our cars along with hooded lights that showed not more than 25 feet of the road in front of the car.

It fell my lot to get together a crew to cut logs for a gypo logger who was to cut timber on Lost Creek. A new bridge had been built about a mile above Batterson and a logging road brought back to Lost and Cook Creeks. One of the first men to come looking for work was Cougar Bill. He said he was expecting his draft call to come just any day. He needed some money. I asked him about his sheep. He said most of them had not lived through the past winter and that a bear had got his last sheep. His dogs had got so hungry that they ate dead salmon and the salmon poison had finished them. It is odd that cats can eat fish but dogs cannot.

Bill always seemed reluctant to tell anybody his last name. I just entered him as Cougar Bill on the time book. Since the bookkeeper lived at Batterson Station just across the river from Bill's cabin, I told Bill he would have to go to the office and write in his name and Social Security number. When I turned in the time book at the week's end, I told the bookkeeper that Bill would be in to sign on the pay books.

Bill came to work on Monday morning but didn't show up for work Tuesday morning. I stopped at the office that evening and the bookkeeper said that Cougar Bill had stopped in to say that his draft call had come and he would like some money.

The bookkeeper could not pay him in full because he had no Social Security number. They finally decided that Bill could have only a draw check and that he could get a final settlement when he got his Social Security number. Bill didn't give his last name. He said, "Just make the check out to Cougar Bill. They'll cash it in Wheeler."

About two months later, a good-looking soldier came to the office and said, "Here is my Social Security number, and now you can pay me in full."

That bookkeeper had known Cougar Bill for ten years, but he looked so different all shaved, with a hair cut.

It was in the spring of 1946 when a car drove up to my house in Rockaway and a well-dressed couple came to the door. I invited them in. The man seemed to know me quite well. He called me "Howard" and introduced the lady as his wife. When he said they had gone up to Lost Creek but there was no trace of his cabin, then I realized that this was Cougar Bill. When I asked him to tell just how old he really was, I learned he was nine years younger than I.

*Whitie Bennett, with his ax and power saw, on the eleven-foot stump of the
450-year old spruce tree he cut at Menefee Logging Co. near Kernville,
Oregon, in 1955. Photo from the author's collection*

Rough-house Dixon

Dixon could do anything needed in a logging camp. He was good at working his men, and
well-liked by them. He also was one of those drinking and fighting men when in town, blowing
his wages.

That summer, the road was built from Corvallis to Alsea, and down that river to the head of
tidewater, and down Alsea Bay to Waldport. Now we had a shortcut out to the Willamette Valley
and Corvallis. No longer was the only way out up the beach, with ferries at Waldport and
Newport.

Dixon and two of his friends went out to the valley and bought a new 1923 Ford Model T car
and drove it over this newly constructed road. From Waldport, the six miles to the camp was on
the beach. Dixon, with too many drinks for driving his new car, let one of the younger men do the
driving. It was real exciting to drive on the smooth beach, after all those miles on a primitive
road. There was a rough road from the beach to the camp.

It was a Sunday afternoon, and low tide. People from camp were out looking for razor clams.
Cars on the beach were not a usual sight. Here came this car, speeding down the beach with men
in the car shouting and waving their hands. They attempted to make the turn into the camp road
too fast. The car skidded, then turned over. It spilled the three men out onto the wet sand. Soon
help was available to upright the car, with little damage to the vehicle. Sliding on the sand did
little damage to the men, but it did help sober them up a bit.

A few years later, Dixon brought in some attention for himself at a Fourth of July
celebration at one of our coastal towns. There was a loggers show. At this show there was the
usual high-climber event, in which a pole is raised and held in place with guy wires. Dixon,
drunk as usual, insisted on taking part in the tree climbing. When he was not allowed to be a
participant, he went out to where he could reach one of the guy wires and started hand-over-
hand climbing the guy wire. When those in charge tried to stop him, he was too high to be
stopped by grabbing a leg and pulling him down. He just kept going hand-over-hand up the
wire, his movements getting slower and slower until finally he couldn't get the next handhold.
He hung for a moment with one hand. Then, as we all watched him dangling up there,

many of us with our mouths open, he let loose and dropped. In coming down, somehow he turned ends, so he hit his head and made a dent in the top of a car that just happened to be parked under the guy wire. Dixon bounced off the car top and lay on the ground for moment, then got up, looked around and asked, "Has anyone a drink?"

The last I heard of Rough-house Dixon was in 1950. He was still rated a good logger and worked at Valsetz, Oregon.

White Hope Anderson, high climber

When I went to work at the Manary Logging Camp near Yachats on the Oregon Coast, it was a camp with three yarder landings. The high climbing was a full time job for one man. Anderson was a good logger and high-climber. He also liked deer hunting. He had his own speeder and often came into camp with a deer. In 1923 there were few people in that coast country and no one cared about game laws.

Anderson was a show-off at times. He did stunts that by today's safety rules would cost him his job. When he topped one of those tall spruce trees, usually at 200 feet, he would slip his safety belt and stand upright on the top, even before it stopped swaying. He was known to even do a head stand up there.

The old-growth fir and spruce trees that we cut were extra tall. I remember getting 240 feet of logs out of one tree. The workman, as he cut the limbs off the spar trees, watched for signs of rot. He usually rejected any tree that showed rot, or cut the top off below the top rot.

I don't think I ever heard any other name for this man than White Hope Anderson. In his younger days he always won any fight he got into. He got to thinking that his future was in the boxing ring rather than topping trees. All the boxing promoters were looking for a white man to take the World Championship from the first non-white man to win that prize. Young Anderson went to Portland and had no problem getting into the fighting game. He was welcomed as a large, strong and young prospect. When young Anderson was put into the ring with an experienced fighter, his hopes were soon knocked out of him. He went to the usual hiring place, paid his fee, and shipped out on the first high climbing job posted on the bulletin board that showed where logging jobs were to be had.

When he showed up in camp with his climbing spurs and safety belt, his friends all started calling him "White Hope."

Buster Brown

When the depression started in 1929, Brown was in charge of a logging camp near Grande Ronde. He had saved his money and bought several houses in Portland. He thought that he could retire and live on the income from his investment. I liked the way he put it. He said, "I just sit there and ate those houses piece by piece." Now he had to start earning again.

This man was one of the best logging superintendents that I worked under. He had many years experience behind him when he came to take charge of the Menefeee logging operation at Siletz, Oregon, in 1944.

I was in charge of the cutting crew, and with this small operation I marked and scaled all the trees cut. It seemed that most rigging men have the mistaken idea that we of the cutting crew have no knowledge of the problems of yarding the logs up to the landing. In the days when the yarding was done with a spar tree, the location of this tree and the space for the landing had much to do with getting out the logs. I soon found that when the rigging men came looking for the best spar pole and landing, one should never take them to the best place first. Take them to less suitable locations, then take them to the best location.

Buster Brown knew that a good cutting crew foreman or bullbuck would have full knowledge of the land. He would know every hump and hollow and any near-flat place. In this steep coast country, any place on the mountainside less steep than a house roof is called a flat. Buster, when a new landing was to be located, would tell me, "Just leave a tree where you think the place would make a good landing. After the trees are all cut, I can better choose the landing and you can cut out the trees that are left."

The author's brother, Wayne Brunson, skilled saw filer, who worked in this shed in the 1940s to keep hand bucking and falling saws sharp. Wayne filed blades for the McCulloch company's power saws as well as for logging camps.
Photo from the author's collection..

Sometimes, if a good spar tree was not in the best place, it was necessary to bring in a spar pole and raise it. (See Chapter Five for how this was done.) Now that we have the mobile yarder with the collapsing spar pole, we no longer need to leave a spar tree or to bring one in and raise it on the landing.

Charlie Nettles, Camp Superintendent

Charlie Nettles, in the early 1930's, was in charge of the Markham and Kallow logging operations on the Little Nehalem River and in the Onion Peak area. Charlie, or "Old Vinegar Face" as the men who worked under him usually called him, was a good logger and ran a very efficient camp. Charlie was a very exacting man. His first interest was getting the logs onto the railroad cars and to the log dump at the head of tidewater. From there they floated down to the lower Nehalem Bay for rafting to the mill on Grays Harbor in Washington. Charlie had little regard for the welfare of the men working for him.

That was in the days when the large timber holders built their own private railroads to haul out the logs. The camp was on the main highway going from Nehalem Bay north. This highway was the 101 route until the road around Neahkahnie Mountain and north to Cannon Beach was built. Markham & Callow's was a well-equipped operation with bunk houses, a cookhouse, and dining hall to accommodate well over a hundred men. The shop area was equipped to take care of all the needs. With this camp located on a good highway, about half the workmen lived within easy driving distance.

For years the men were hauled from camp to workplaces on open flatcars. In the wintertime this was a real hardship on the men. Rain, snow or blow, the crew were hauled to work in the morning. No matter how hard the wind blew in rain or snow, the men were held on the job until quitting time. Then they were hauled back down the mountain to camp all wet and cold on that open flatcar. Men would arrive at camp having to drive their cars as much as twenty miles home, wet and so cold that they needed help to get out of their work clothes. When men complained, he would say, "Go to the commissary and buy heavy wool underwear."

The author trimming root flares from the butt of Whitie Bennett's 11-foot high log so it could be hauled by truck. Note the end of the "lightweight" 70-inch saw's blade sticking out of the log at right. Photo by Dick Osborne.

In 1934 the Woodworkers' Union was the first organized. Then, and only at the demand of our united group, was a closed boxcar provided for hauling the crews out to work and back.

The logs from this camp were made into ocean-going rafts and towed to the Grays Harbor mill. They lost so many logs in the winter storms that they gave up on this method. The logs were then loaded on the Southern Pacific railroad and shipped by rail. When the State Highway was paved all the way to Astoria on the Columbia River, this company, with other logging companies, started hauling their logs to log dumps at the mouth of the Columbia River.

We moved to Rockaway, in Tillamook County, on the Oregon coast, in the fall of 1930. Work was so intermittent for the next 10 years that I had ample time to fish every stream and hunt every mountain ridge in the whole of Tillamook County. The coastal mountains were a vast timbered area with only a few primitive roads. The fire lookout stations on the higher mountain peaks were reached only by trails.

The best areas, and hardest to get into, were on Cook and Lost Creeks. These streams flowed into the Nehalem River a few miles upstream from the Hammond logging camp at Foss. Cook Creek, opposite Batterson Station and Lost Creek, was about a mile upstream. The county road ended a short distance above Batterson Station. At that time the Southern Pacific Railroad came down the north side of Nehalem River. The logging company's railroad crossed the river at the camp and came up to Cook Creek, and up that stream about two miles. Hammond's was a large operation, with a cookhouse that could feed 250 men. They shut down that year, 1930, due to the poor demand for lumber.

Bill Harlan, Logger!

He was one of those capable workmen who work their way up to the top. As a young man, working for Hammond Logging Company, he could do any work needed to run a logging operation. He was in charge of Hammond's logging when they closed their operation at Mill City and built the camp at Foss on the Nehalem River.

The Foss Camp employed about 300 men. The cookhouse there could seat 250 men and there were more than twenty family houses. Bill Harlan, when I first met him, was in charge of this logging camp. It was a railroad logging camp and needed four locomotives. They used at least three Ledgerwood yarders. They had about twenty miles of railroads. Then times changed, and they went down to one yarder and one steam locomotive.

With the devastation by fire in the area called the Tillamook Burn, and the poor market for logs, the Hammond Company gave up on logging and sold off all their equipment. By 1940, all the rails were taken up and the logs were hauled by log truck. Most logging then was being done by small operators with only one yarder.

Bill Harlan had bought a small tract of land where he built his own house, thinking it would be his last move as he would soon reach retirement age. He was a good manager of workmen. All who worked for him liked him well.

In 1952, I took over the cutting crew for the Menefeee Logging company on Lost Creek. Good men were hard to find, with the war drafting most of the younger men. I was surprised one day when Bill Harlan came to me and said, "Howard, I need to go to work. I am not ready to retire. I once bucked logs. I can still run a bucking saw."

Our crummy came right by his home. I told him, "Be on the crummy with my cutting crew. I'll tell the driver to stop for you."

Since my men were paid by the number of board feet they cut, Bill was well pleased with the job. When it came to getting out logs, it was the older, experienced men's game. Because of the war, and the scarcity of good men, the pay was upped. In fact, the pay he made on my cutting crew was more than he had been paid to superintend the Foss Logging Camp.

Bill Harlan had started his logging career at cutting logs and worked his way up to managing a 300-man logging camp. And his last work was back to cutting logs with a handsaw. By the time power saws came to the woods, he was ready to retire.

I remember one hunting season when he knew that several of the men were hunting on that last weekend of the season. At least three miles up the railroad to the head of Anderson Creek was where the best deer hunting would be found. In the afternoon, he took a speeder and went up to the end of the railroad. He waited until he was sure that all the hunters were out to the railroad track with their deer so that he could haul them down to camp.

Dave Burwell

We have heard about the man who started at the head of an operation and then worked down. Dave was a good example. When he first came to me looking for work he had just graduated from the state university in forestry. He had his degree in book learning, but no practical experience. He had just signed up for officer's training for our armed services. It would be nearly two months before he would be called to start this training. He needed to earn money to live on, and he also had some serious courting to do with the lady of his choice. His friend, brother to his girl, was working for me. Norman was a fair timber faller. I told Dave that he could go and work as second faller with Norman. Dave worked about six weeks, until his call came to enter the Officer's Training School. In due time, he went into the U.S. Navy. By the end of the war he had been promoted up, until at the time of his discharge he had the rank of Lieutenant Commander.

In August of 1950, Dave called me. The company he worked for was taking over the Warner logging holdings in the Siletz country in Lincoln County, with headquarters at Cutler City, Oregon. It would be only a one-landing operation. The crews would be transported from Cutler City to the

By the 1940s, trains for hauling logs from the woods had followed ox teams and teams of horses into history. Logs now traveled on rubber-tired trucks. Here men use a crotch line suspended from a rack or boom that swivels on a spar tree to load the top log of three on Richard Kendall's log truck. Photo courtesy of Charles V. Ames.

landing by company-owned buses. Dave would have charge of this logging op eration. By his having earned rapid promotion, Dave had proven his leadership capabilities. He had graduated with good grades from the Forestry Department at Oregon State University.
We who had spent many years working in the logging industry wondered how he, with less than two months of practical experience, could manage a crew of experienced loggers.

He asked me to come and take charge of the cutting crew. Where we would start working was in 80 acres of old-growth fir. Most of it was on almost level ground with only a small area of steep ground. He also hired a good tractor operator and a good rigging man who knew how to do all the work needed to get out the logs.

The first problem was that Dave tried to treat his loggers much like he had treated the men under him in the Navy. He bought good machinery, but could not agree on the wages to be paid the men. The cutting crew would work at piece work. I knew from many years of experience figuring the price per thousand board feet just what it would cost to get this kind of timber cut. Dave would have his way and set his price. Good timber cutters would come and work, but by the third day they would leave, saying they would not work that cheap.

Finally, in order to hold a cutting crew, he raised the wages to the price I had suggested we start at. Then the bookkeeper had to apply the new wage scale, and send checks to all the men who had come and worked just a day or two. Some men who had to depend on this work had stayed on the job until the working wage was established.

After a few months of trying to do work done as it should be done, I also was ready to look for work without so many problems. Mostly due to Dave's inexperience, the men not staying on the job, and the logs not coming from the landing, the management had to let him go. Dave may have got this job due to a relative owning an interest in this logging company. It was midwinter, and Dave had a wife and children to feed and no job. He finally got work setting chokers behind a Caterpillar tractor east of Eugene, Oregon. When you can log with a tractor in the winter time it is a muddy job.

I will have to give Dave credit. In time, and with practical experience, he earned the position of head forester for one of the larger timber companies in our state. He also told a person who would tell me, "That Howard Brunson sure knew how to cut timber."

Conley Poe

Conley Poe was a second-generation logger. He was an unusual logger in that he never smoked, drank hard liquor, or used profane language and yet was a very rough man. His wife, in contrast, was a faithful church woman. In spite of their character differences, they lived happily many years. They had two sons, who also started working in the logging at an early age.

I first met them in 1932. They lived at the old camp a few miles north of Mohler. They came with some friends to the church at Rockaway. December of 1932 was an unusually wet month at Rockaway. We had 36 inches of rainfall in the first 22 days. We had planned on spending Christmas with Olive's folks at Monmouth. The Poes planned on spending Christmas with Ruth's parents at Falls City. The Poes did not have a car they could drive at that time. They went with us in our car.

This made us a load of four grownup people, our three children and their two. I remember that the road between Rockaway and Barview was closed by a slide at Smith Lake, due to a storm. We had to take the old plank road on the west side of the lake. We were fortunate to get through, as the lake water was up on the old road so that the planks were floating. However, the storm was over and the sun shone the five days that we spent out in the Willamette Valley. We always referred to a visit of this kind as a trip to "The Valley".

Conley Poe was a good man in the woods. He was a good timber faller. He preferred to fall the trees rather than buck them into logs, chopping right-handed. Most of the trees on the steep hillsides of our Oregon Coast are so timber-bound that in the days before the power saw, we had to chop the undercut with axes. We used five-pound, specially-designed axes with 42-inch handles. In chopping the undercut in those large trees, we put a bend in those long handles, so that we could reach and chop out the middle and not bang our hands. We had a problem reaching into the center of a large tree up to 10 or 11 feet in diameter. I remember spending as long as four hours chopping on the undercut, and then as much as two hours with a ten-foot falling saw in the sawing of one tree.

When power saws were developed, one man could fall a ten-foot tree in less than one hour's time. In the hand-falling days, we would "board up" as much as two boards high and work on the springboards. The springboard was a plank six to seven inches wide and about five feet long with a circular-shaped iron on one end that fitted into a notch cut into the side of the tree. When properly placed, we could, with the right foot motion, move this springboard about a quarter circle. A large tree might require one hole for cutting the undercut and another to start sawing on the back of the tree. Sometimes, due to the size or shape of a large tree, we would have to put in three springboard holes. This all took time. Two men could spend most of a day just falling one tree. When the power saw came into use, one man standing on the ground could cut an eight-foot tree in less than an hour, and as quickly as in twenty minutes.

Conley and I worked for several years in the 1930s and 1940s in the same logging camps. Since we were both right-handed choppers, we never cut trees as working partners. While I was a good tree faller, I also was good at bucking. Often I was the man bucking the trees Conley felled. Conley liked to work in the camps that paid by how much work he did. We loggers called this "working by the bushel" or piecework.

There were two systems of working the falling saw by hand. Some of the larger men pulled a heavy saw, in that they held it to the wood. I preferred to pull a light fast-stroke with a swing to the natural bend of the falling saw. I remember one time when the left-handed faller didn't show up, they asked me to fill in at falling that day. Not all men who are good with the bucking saw are as good with the falling saw, as it is a different type of saw. You have to use a stroke suited to its shape. I knew that Conley used a fast, light stroke. When we had chopped the undercut and were ready to start sawing the back, I started out with a medium-fast stroke. Then as my partner met my pace, I kept going faster, until finally Conley called for a rest. I smiled at the way he indicated we were going too fast. He got down off his springboard and looked at the saw cut. I came back to see just what he had stopped work to see. He asked me, "Did you see smoke coming out of the saw cut?" It was his way of telling me to slow down the saw stroke a bit.

Conley all his life worked at cutting trees.

We had it tough during the Depression years of the 1930s. Wages were cheap when you could find work. It seemed that the logging camps only worked a short time. About the time we got caught up on our debts, they would close down for several months.

It was with Conley Poe that I learned how to hunt deer and elk. For several years, it was only with the clams in the bays and the wild meat that we managed to feed our families through the winters. We always tried during the hunting season to get in enough deer and elk meat to last us until the next season.

Conley had a special desire for hunting elk. I remember when the first elk hunting season was announced in Clatsop County. Conley went in advance and located a small elk herd with a five-point bull. He went with a companion and laid out just over the ridge all night. When daylight came, he, with one shot from his 30.06 Savage rifle, laid that big bull elk on the ground. After cutting its throat we got it on its back, cleaned out its insides, then skinned it, keeping the meat on the hide. Then we cut it up and put the meat in white meat sacks. That big Conley would take a quarter, weighing near 170 pounds, in one pack. Ordinary men could put only 100 pounds in their packs. Conley would make the half-mile to the car out on the road without a rest.

Conley liked to shoot his deer or elk in the neck. I always say, "Shoot it in the front shoulder." I see too many shots that miss the head or neck.

I will have to tell of one time that his neck shooting didn't pay off.

We were badly in need of meat and both out of money. He had only three shells for his 30.06 rifle. He came to me to see if I had enough gas in my car to take us to where he knew some elk were feeding. I had an old 32.40 rifle that had seen better days. It shot fairly straight but the bullets, after leaving the barrel, would turn end over end. When this old gun came to me, it still had a few shells in the magazine. I took this old gun with me.

Well, sure enough, we slipped up behind a large stump and looked into a little canyon, and here were six elk. Conley, with his first neck shot, just clipped he hair off the top of the elk's neck. Three of those elk ran out of that little canyon, with Conley just clipping hair. I kept telling him, "Shoot them in the front shoulder!"

With his ammunition all used up, he took the old 32.40 and clipped another of the animals. Finally, as the last elk ran out, he shot it in the front shoulder. We could see the hair puff up as that bullet from the old gun hit flat-ways. When we dressed out the animal, we found the bullet lodged in the hide, after going through the front shoulder, and it was still sideways. That old elk was the toughest meat we ever packed out of the woods.

When the Poes moved back to Falls City, to the house Ruth inherited from her parents, he started working out of that place, and I lost a good hunting partner.

We didn't see the Poes very often for quite a number of years. After we moved to Monmouth in May of 1980, we were able to renew old friendships. Conley is gone now, and Ruth died in 1969.

Eccles and Annie McCaw

When I remember Eccles and Annie McCaw, I think of them as an example of the partnership of marriage. These two people worked, played, and worshipped together. They graduated in the same class at the high school in Dayton, Washington, and were married that same day, right after the graduation service. They were active members of a church and in community activities in the many places where they lived. Eccles would try any work that was available and was always a good workman.

His first introduction to working in our forests was cutting firewood near Pedee in Polk County, Oregon. He had acquired a Chevrolet truck and came to Rockaway on the Oregon coast to haul logs for a small gypo logging company. He found that his truck was too light for hauling the logs and traded that truck in as down payment on a larger truck. This was in the summer of 1932. The price received for the logs was too cheap for the gypo logger to operate. When the logging company couldn't pay Eccles McCaw, he lost his truck. It left him with a wife and two children and no money to pay the bills.

Eccles was a fair hand with a falling saw. He could chop either right- or left-handed. That winter, we felled timber together for another gypo logger. When spring came, that logger could not pay us even the minimum wage of 40 cents an hour. At some previous time, the McCaws had learned to pick apples, so when apple picking time came, Eccles gave up on tree cutting. He and his wife, Annie, went to Hood River and picked apples.

In the next few years, the McCaws did several different things to make a living. With the start of World War II, they found that salvaging scrap iron paid good money. Then, again trouble caught up to them. Their truck that they hauled the scrap iron in broke down.

At that time they were near Gilcrest, Oregon. Learning that the logging company there needed cutters, Eccles went to work on the cutting crew. This was the start of another period of prosperity. Eccles, in his spare time, got started cutting poles. In that area there were numerous stands of small pine trees just the right size for power line poles or pilings. The McCaws also built themselves a house and started a restaurant. Annie was as adept at driving nails as she was at cooking in their restaurant. She could also operate the engine when loading out a carload of poles.

The McCaws loved to fish and hunt. Although Annie had only one good eye, she always shot her buck deer and never hesitated to dress out her own kill. While they often lived and worked in other parts of our country, they always came back to Central Oregon for the good fishing and hunting.

The year 1949 was a bad year for the McCaws. They tried to operate a small sawmill. The log market and the lumber market were depressed, and again the McCaws lost everything. They moved to Florence, Oregon, and lived in a shack of a place. Again, Eccles started falling timber. It happened that again I needed a left-handed partner, only this time I taught him to fall trees with a power saw.

Eventually, the McCaws got into the logging again. They managed to get an old tractor that he was able to repair. They bought small farms that had some trees to be logged. Again they were having prosperous times. Then they bought a larger ranch where the timber was not as good as represented, and the log market went into a slump. Again, the McCaws found their cupboard bare.

The next I heard of them, they were logging in New Mexico. In 1961, Eccles came to Oregon and again worked with me. By now, Eccles was a very good hand at cutting trees. Annie was a fine cook and often cooked in restaurants. They had their own restaurant in Dallas, Oregon, but again they drifted back to Bend and Central Oregon. Eccles could do any work in the logging industry and could always find work. They bought some land east of Bend and this became their permanent home.

It was not until retirement time came that they found their true niche in their life plan. It started when they found that they could sell. They were natural sales persons. These no doubt were the best years of their lives together. They started with selling a bottle cutter. Then they got into collecting and selling barbed wire. They would get permission from some of the ranchers to remove all the tangled barbed wire from their ranches, untangle and sell it to someone else. They were able to spend winters in the south, then come home to Bend for the fishing and hunting. They found several items to promote and sell at a profit, like weathered barn siding boards for picture frames. Annie, with her cooking expertise, could always generate extra income when needed by cooking in a restaurant. We have to admire these people, so capable they always found work.

Joe Flora

Joe Flora, as a logging camp superintendent, always tried to get the logs out faster than any other camp. He earned his reputation as a highball logger while in charge of the camps that logged for the Portland Lumber Company. This company's timber holding was part of that vast stand of old-growth trees on the west side of the Willamette and Columbia Rivers that extended from Portland down the Willamette River to Astoria and west to the ocean.

Several railroads were built from the rivers westward to haul the logs out to where they could be dumped into the water, rafted, and towed to the mills. These railroads in the 1920's served camps noted for having the fastest and best loggers. I heard the men boast with pride that they had worked at one of this company's camps. They had left because the operators, in their hurry to get out the logs, had little regard for the safety of the workmen. The camps operated by Joe Flora had a very bad reputation in regard to safe working practices.

I first worked for the Flora Logging Company in 1929. At that time this company had three camps west of Carlton, Oregon. Their railroad went up the north fork of the Yamhill River and over toward the Trask River. The headquarters camp for the railroad crews and two camps that housed the loggers each had two yarders. Each camp had about 100 men working.

The fall of 1929 was the beginning of the Depression years. The logging camps were shutting down due to falling market prices. My partner and I went to Portland to look for work. The only work listed on the bulletin boards at the hiring offices was for choker men for the Flora camps out of Carlton. I was not anxious to work on the rigging. The job notice stated that we would only work a week, then the camp would shut down until March. The Flora camps were up near the summit of the coast mountains at an elevation that got snow too deep for logging. They always closed from December 20th to March 1st in the spring.

Neither of us had ever worked for Joe Flora's camps. My partner said, "How can you claim to be a good logger without being able to say that you have worked for this man?"

We went to Carlton, checked into the office and were told to go to where the logs were dumped and ride the train up to the first camp. To our surprise, the train engine was behind the empty log cars and was pushing them. We went up the Yamhill River several miles, then the train stopped for a moment, then went forward and started climbing a steep grade. As we left the river we could see the railroad tracks below us and realized that, instead of making a curve to get up the mountain, the railroad made a switchback, or "y." Now the engine was in front. We soon stopped on a siding to let a loaded train pass. We were surprised to see the length of the logs on it. They seemed to be about 80 feet in length with a top diameter of 30 inches. We were four hours getting from the log dump at the mill pond up to the camp.

I was a bit disappointed at the Flora camp and equipment. I had worked at Manary's camp where everything was new. Here at Flora's camp everything had been in use many years and looked to have seen hard use. The yarders were powerful Seattles, built by the Washington Iron Works and named after that city. The mainline was equipped to carry two butt riggings and two of those heavy chokers that require three men to handle them. The butt rigging is the device on the mainline choker cable that, after it is put around the log, the knob on the end of the choker is slipped into.

We had only three choker men, so we only used one choker. It was surprising, the power of those 12x14 steam engines. When we had set the choker cable around the 80-foot log with a top diameter of less than 30 inches, we gave the go ahead signal. That log would just seem to fly. If we saw that it would hit a stump we would have to signal, "Stop." If the log hit the stump, the choker cable would sometimes cut the end off the log. Then we would have to go again and put the cable around what was left of the log. Even though we had only three choker setters, one day we yarded and loaded out 17 railcars. Each car would carry more than 10,000 board feet of logs.

If I remember right, our wages were less than six dollars a day. There were men on that logging crew who had stayed in camp since the Fourth of July two-week shutdown. Some had saved their money and would live well in town until the camp would again start up in March. The work week was six days, but there was always work to be done on Sunday for anyone who cared to work that day. Our rigging slinger said he usually worked three weeks, every day, then took off three days to go home for a short stay with his wife and children.

In a camp out in the forest like this one, card playing in the evening was the usual pastime. There would be several games going, some just for pastime, others playing penny ante, then there would be more serious gambling for larger stakes. There was the big game where it took at least a hundred dollars to get into the game. Bets in this game could up into the hundreds of dollars. I often wondered how some of the workmen could, at their wages, sit in this game and lose their hard-earned money.

Some men made it a practice to work this high-stake game. They usually had one of the easier jobs about camp. At this Flora camp, it was the pump man. All the yarders and loaders were steam engines, fueled by crude oil. Each landing would have a pump that supplied water. The pump man had his own speeder on the railroad. When a landing signaled by whistle that they needed water, the pump man went to the usually gasoline-powered pump that supplied the water and pumped water until the whistle signal indicated that their water tanks were filled and he should shut off the pump.

We worked until the end of the week and shut down time. At that elevation, and with cold weather soon to come, it was necessary to drain all the water from everything. The engines at the landing shut off the oil and put out their fires. Then they opened a valve and drained out the boilers. The water would be mostly steam. It was quite a sight as we came into the landing to see that large cloud of steam as it hissed out of the boilers and floated across the mountain. Two clouds rose from our two landings, and in the distance we could see clouds of steam as they blew from the machines at the other camps.

The forest fire in 1933 completely wiped out the Flora logging operations. Everything made of wood burned, all the buildings and railroad trestles and even the large logs that the yarder and loader engines were mounted on. The donkey sleds, as they were called, burned because their fuel oil had soaked the wood. The spar poles, fully rigged, were left standing.

I worked for Joe Flora again when he came to Foss, Oregon. He logged some timber for the Hammond Company and used their equipment. He had only one yarder and one tractor. He first

Bushes regrown a few years after a forest fire provide browse for deer. Boulders in the stream give fish shade. Photo by the author.

yarded a unit on Cook Creek on the railroad, then up a smaller creek to the ridge top from which he hauled the logs down the mountain on log trucks. By now he was very safety conscious. He would tell us, "Get the work done as quickly as you can, but don't get hurt."

Dan Valley, Bullbuck

As a logging company foreman, he knew how to get the timber cut, but was never inclined to give all the men in his cutting crews a fair deal. I first heard of him when he had charge of the log cutting at a camp on Buck Creek. This logging operation was on one of the railroads that logged the country west of the Columbia River and over into the bend of the Nehalem River, into the finest stand of old-growth fir in the State of Oregon.

In October of 1926 I decided to find work in the pine timber of central Oregon. I went to the camps of the Shevilin-Hixon Company south of Bend, Oregon. They had one camp on the highway, 30 miles south of and near Lava Butte. The other camp was up in the hills, four miles east of the highway. I wanted to get on the cutting crew. I was told that I could work on the section crew just moving this camp into its new location, deeper into the forest from the area they had already logged off. All the buildings were on wheels or so constructed that they could be moved on the railroad.

At this time all the logs were moved from the forest to the mill by railroads. The main street of this camp was a railroad with short spurs branching out for the buildings. It was interesting work, moving this camp and placing the buildings in their natural order.

In due time, the chance to go to work cutting pine trees came my way. Dan Valley came by where I was working on the crew maintaining the railroads from camp out to where the yarder machines were working. What I had heard about his tactic of giving some of the cutting crews the best timber to cut was apparently true. We were expected to drive our own cars from camp out to where we would be cutting the trees. We could drive our cars almost any place through the pine forest. It seemed that my partner and I would always be started on the poorest trees. Then when we got up to the good trees, we would be moved to another stand of poor trees for making money. We were paid 90 cents a thousand board feet of logs by the company's system of log scale. We had to cut the trees low to the ground, limb the trees, and cut the logs into 32-foot lengths with an 8- to 12-inch trim. It was rumored that if we paid a tribute in cash to the bullbuck, Dan Valley, we would be allowed to cut those trees we could cut faster than the trees that required more time limbing than log cutting.

Dan Valley was never known to be contented with the wages the company paid him for his overseeing the cutting of the logs. He was accused of always finding ways to extract money from the men. The company furnished houses for their foremen. Dan Valley always had a "housekeeper" and a new car. At all times he wore good clothes and kid gloves.

I don't know how long Dan held this position with the Shevilin and Hixon Company. When the camp shut down for the two weeks on July Fourth, I was well tired of cutting low stumps and tough limbs. Try our best, we could only average six dollars a day. We found we made the most money cutting trees that averaged about 1,000 feet in log scale and grew close enough together that the limbs were small and easy to chop off. When the vacation time came, my wife and I decided to visit our parents at Newberg. As we were expecting our second baby, we decided to stay where we could be near our family doctor who had known us since we were small children.

Well, all good things must come to an end, even for men like Dan Valley. The next time I heard of him, he was cutting trees for a small gypo logger in the Luckiamute country and living in a tent. With the organizing of the logging camps into a union, wages and working conditions were stabilized and foremen like Dan Valley no longer were tolerated in the logging industry. Workmen were allowed a fair wage.

We men working at piecework often ran into this problem: If we worked too hard, or had a fortunate break that brought our earnings up above our average, the management would object when pay time came. Once we cut some trees for a resort owner. A winter storm had blown down some trees. One fell right through the center of one building. They also wanted other trees cut to clear space for more buildings. I had worked a day by myself right after the storm, then when the crew helped me, we worked for two 10-hour days. When I turned in this report with my monthly report, I failed to note the extra 16 hours time. The resort owner thought we workmen were not entitled to earn that much money for a day's work. I then made out a separate report for this work, and made other changes, and included the actual time we worked in hours. The new report actually made us more earnings than the original report, yet he paid on the report that showed our actual working time on the job.

The Cook Brothers

In the spring of 1938 I learned that the Coats Logging Company was going to log a large tract of timber in the Tillamook Burn. They had built a road up from the Wilson River and would be working some long gentle slopes between the Wilson and Trask rivers. It seemed a good place to work for the summer, but would be at an elevation of deep snow in the winter. I went out to Claude Evans' house at their main camp to ask about a job. He said he could use one left-handed timber faller and three buckers. I would rather buck logs than have to work and chop as the left-handed tree faller. I agreed to take one of the bucking jobs.

We visited a bit as we were old acquaintances. An old beat-up, makeshift kind of Ford car drove up with Arkansas license plates. Quite a number of men came from the Midwest and the dust bowl areas in the 1930's, and later learned to be good loggers.

Three very tall, slim men kind of unfolded themselves, two from the front seat and one from the back. We wondered how such a tall man could get into the cramped back seat. In that day some cars had what was called a "rumble seat" in back of a one-seated vehicle. As these three men approached the house, Claude said, "It looks as if here comes the balance of my cutting crew."

Walter, the tallest one, acted as spokesman. Their name was Cook. They said they could use a chopping axe and could cut wood with a bucking saw. Steve, the middle brother, said that his natural way to chop was left-handed. So it was decided that Steve would take the felling job and Walter and Lester, the youngest, would try the log bucking. We would be working on gentle slopes, and they would have experienced men working nearby who would help them with any problem that gave them trouble due to their inexperience. We were all surprised at how these three men made out at this work that was new to them. Lester worked right alongside of me and, in just a few days, he was cutting as many logs as I could cut each day.

A logging camp of the 1940s. Bunk houses were long and narrow so they could be moved on the logging company's railroad to a new site when nearby trees were all logged off.
Photo courtesy of Tacoma Public Library, the Richards collection

The three Cook brothers had only worked for two paydays when they got word that their father was very sick and would not last many more days. Those boys got leave to be away from their jobs while they went home to the hill farm of their parents. The three of them got into that old car and drove, taking turns at the wheel, and stopping only when the car needed fuel. They drove straight through, and arrived in time. As soon as the funeral was over, they drove that old car back to Oregon and only lost eight days from work. They all three took the trouble to get married or make arrangements for their sweethearts to come to Oregon just as soon as they could afford to rent or buy a house. These three men, with time and experience, became good timber cutters.

CHAPTER THREE

MORE LOGGERS

Preacher James

Lloyd James was an ex-prizefighter. He had fought most of the best heavyweight professional fighters around. In his late twenties he realized that if he continued he would be as punch-drunk as some of the older fighters no one wanted to fight with anymore.

He married his sweetheart, the girl who had lived next door as he grew up in a poor community in Oklahoma. Until he took up professional boxing, he had never lived in a house with electricity and indoor plumbing. He came west to the Northwest Christian College in Eugene, Oregon, to study to be a preacher. He had been ordained but he never preached as a regular pastor. He ran out of money before he could graduate. He had heard that the quickest way to earn a stake was by working in the logging camps.

It so happened that at our little church at Rockaway, Oregon, the pastor was a student at the church college in Eugene. This man advised James to come to Rockaway and talk to one of the members of our church. James came to Rockaway in May of 1939 with his wife and three small children. They rented a house, then he came to see me about how to learn to cut logs. I still had a few days of work to do bucking the logs that the Palmer brothers fell for Oney McDonald and Bart McKinney in a cutting where we had agreed to wait for our pay until the logs were sold. There was no reason, since I had extra tools, that James could not work with me and learn.

James was a large, well-built man but unused to the woods. I had worked with him only three days when I got the chance to go to work at the Markam & Callow camp. This company had finished their railroad logging and, under new management, were logging their holdings on Sweet Home Creek, a branch of the Little Nehalem River. The ground was even enough for tractor logging, and the logs would be hauled to the Columbia River at Astoria by truck. James could finish the log bucking for the gypo.

I had worked at the new job just a few days when I had the misfortune to have a sharp axe bounce off a hard knot and cut my foot. I would be off from work for a few days. This gave James an opportunity. He went up to the camp and the bullbuck took a liking to him and gave him a chance. James was a likable fellow and the crew all helped him. It took a lot of cuts and bruises, but with experience James became a good cutting crew man. When the fellows learned that he was an ordained minister they all called him "Preacher."

Cuts from saws, especially hemlock and cedar, can be painful. The sap that comes with cutting green trees will cause an infection. Preacher James cut a finger, and I blame the old doctor at Wheeler for not taking better care. The finger had to be amputated to stop the infection. It had been almost four years since James and his wife had been back to Oklahoma to visit their parents. With him not able to work while the finger healed, it seemed a good time to make the trip, excepting for money. Those resourceful people planned that out. The check for the loss of the finger would pay the train fare. Preacher James could go half-fare, his wife full-fare, and only one of the children was old enough to be charged. There would be little or no money for food. They had a solution for that problem. They wrote relatives and made arrangements for overnight stops where the relatives would keep them overnight then provide enough food to last them until the next overnight stop. They were able to make an overnight stop with relatives who took them through the Carlsbad Caverns. With all this good planning, the James family was able to take a two-week vacation trip, visiting relatives.

The Tillamook County Fair is the big event of the year in our coastal county. At fair time, the carnival section usually has a fighter who offers any local fighter money for each round he can last with him. It is usually an ex-pug who is no longer good enough to get well-paying engagements. At the 1939 fair the carnival had such a fighter. They offered $50 a round for anyone who would get into the ring. Very few men could go the full three rounds of the fight's limit. Some of our cutting crew lived in or near Tillamook. These men decided to get Preacher James to meet the carnival fighter. They persuaded Preacher to give it a try. After all, $50 was more than three days earnings at bucking logs.

The fight was set up for Saturday evening and advertised as a Tillamook Logger against the Carnival Fighter. James, after talking with the man, agreed that they could put on a good three-round exhibition. The fight attracted a full house, or full tent. The fighters were introduced, first the carnival fighter, then Logger James. James came to the ring wearing a hickory shirt and staged-off blue work pants, and wearing his caulk shoes.

Now here was a problem. The referee refused to let the fight start until the logger took off his caulk shoes. James refused to fight in his stocking feet. Finally another large man, who had come to see the fight wearing tennis shoes, was willing to lend them so the fight could get started. With the tennis shoes on James's feet instead of the caulk boots, the fight got under way.

James, in spite of not having had on a pair of boxing gloves for years, gave a good exhibition of ring footwork and punches and counter-punches well into the final round. As the end of the third round neared, the carnival fighter thought he saw a chance to get in a hard punch that could well end the fight and save some money. To his surprise, he was jolted with a punch that nearly floored him. When the bell rang that ended the fight, the spectators all shouted, wanting the fight to go another round.

I doubt if Preacher James ever took on another fight. He never took on a regular preaching assignment, either. He could make good wages in the logging camps and work in the woods until retirement time.

Both James and his wife came from large families. Lloyd seemed content with two sons and a daughter. His wife thought that, now that he was making good wages, they should have a larger family. One day while visiting with my wife she expressed the idea that she would catch him in a weak moment and have more children. In due time, two more children were added to their family.

Jesse Bell

Jesse Bell moved to western Oregon in the 1930s, another victim of the Midwest drought. When the Bell family moved to Twin Rocks on the Oregon coast, work was hard to find. Jesse would try any kind of work that would help to put food on the table. He first cut wood, and this led him to learn the logging business. Since he was good with machinery, he managed to obtain a used Caterpillar tractor. Soon he started logging small tracts of timber.

In the spring of 1942, I teamed up with Jesse Bell to log a unit of timber on the Miami River in Tillamook County. I was to do the falling and bucking while he did the yarding and loading. I, with a partner, felled the trees. Jesse overhauled the motor of this old tractor, and by the time the tractor was ready for work, we had felled all the trees and cut them into logs. In his hurry to put the tractor to work, Jesse never replaced the cab and safety guards that extend from the front of the tractor, up over the cab to protect the operator. I noted this neglect, but Jesse said he could work without them.

I helped get the log landing ready. I had cut several tall slender alder trees to clear the log landing space. Jesse decided to use the tractor blade to push these cull trees into and across the river. All went well, and he had the landing space cleared, but he didn't want some of the alder in the water. He started to shove them all up on the opposite river bank. In lifting with the blade, one tree's butt-end hung up in the river. Then, as he gave the final push, the main part of the tree slipped up over the top of the tractor and pinned Jesse to the back of the tractor seat with such force that it knocked him out. On later examination, we found it broke several ribs on his left side.

I grabbed an ax and rushed out to the tractor. Since there was still a strain on the alder tree, it took only a few well placed strokes to chop it apart. With the pressure released from his body, Jesse was able to get his breath and tell me which lever to use to back the tractor out of the water. I managed to get the injured man off the tractor and into the back seat of his car. Then I drove him to the hospital in Tillamook.

It so happened that the Bells had a 14-year-old son, Charles, who was just waiting to operate that tractor. We had that logging job well-near finished before Jesse Bell was able to come back to work. By that time, the son was nearly as good a logger as his father.

Dale Jones of the Jones Logging Company

I first cut timber for the Jones Logging Company on the Oregon Coast in November of 1950. While this logging company was using his name, Dale Jones only owned a small interest in the company. Dale was a good logger. He could operate and maintain any logging equipment the company owned or used. He was a fair man and got along well with the men who worked for him. The men who were major shareholders seemed more interested in making a profit. That is their privilege and their reason for operating the business. The Jones Logging Company bought and logged timberland and logged tracts of timber for other companies.

They bought a tract of timber on Siletz Bay. Quite a lot of this timber was old-growth fir and spruce. Much of it was young-growth spruce and hemlock. I learned that, while they had three men cutting one unit, they were looking for someone to cut another unit. I talked to Dale Jones one morning and he showed me the timber he wanted cut. There were several large, old, leaning

Faller Cliff Williams using hydraulic wedges to tip a cut fir tree off its stump in 1949, near Kernville, Oregon. Photo from the author's collection.

trees. He said, "Just fall them the way they lean, and don't worry if they break up."

I didn't say anything, but I know how to save timber in falling. The first day I felled three of those threes. He was really pleased when he checked that evening to see them on the ground and each in one piece.

With some of the other large, leaning trees, we waited until the wind was strong enough to make them fall in a way that would save their entire length. There was a lot of trouble with logging the unit his other cutters had worked. When they yarded out our unit, they found that every log had been cut off so that it came right up to the landing. In the other unit, they would find logs uncut. Then they would have to hold up the rigging and get a power saw and make those cuts.

All we had was a verbal agreement to cut the timber. I had charge of the cutting of the logs for this company for seven years. The Jones Logging Co. enlarged their operations until they were logging three landings. They contracted to log for the Longview Fiber Company and for Crown Zellerbach.

When I cut timber. I always consulted that man who owned or had charge of the timber. I cut it the way he wanted it done. After my having charge of the cutting for seven years, someone in the head office decided that I was making more money than a working man should be paid. Since we had nothing in writing, it was no problem to terminate our agreement. I was the oldest member of the cutting crew. There was no reason why I could not stay on as a regular member of the cutting crew. By the agreement, I was to be paid a final settlement as the logs were scaled after being loaded on the log trucks. It seemed the thing for me to do was to stay until the final settlement.

I had furnished all the tools that the cutting crew used and furnished the transportation from their homes to their work. The new arrangement was that the men would furnish their own tools and transportation. The company was to pay each man according to the log scale they cut. The company was to hire the bullbuck, or foreman. A good foreman will carefully look over the timber before he sends his men to work. He sees all the problems, and that the men have the proper tools needed to do a good job. When it is left to the men to furnish their own power saws and other needed equipment, there is the tendency to try to do the work with the least money spent on tools.

This bullbuck Dale Jones hired to replace me may have been a good workman in his day. He was a large, heavy man who never got off the road if at all possible. He just stood on the road and told his men where to work. He had men working against the natural order. This slowed up the work, and made it harder for the workmen, at less pay. I am sure that men received on-the-job injuries doing work that should have been better supervised. A man will work dangerously rather than disagree with his boss. When the men working under my supervision turned in the log scale for a unit, we were sure that the scale would hold up to the log truck scale. Not so under the new foreman. Most of the men who had worked for me all those years soon drifted on to other work and new men were hired.

I worked two more years, but soon after the final settlement for the logs cut under my supervision I, too, left for other work. For some reason, the Jones Logging Co., in the years to follow, no longer contracted to log for other timber owners, and there were no longer small tracts of timber to buy and log. The Jones Company went out of the logging business.

Who ever thought of cutting logs by the pound? Crown Zellerbach first started buying pulp logs by the pound. It was quicker to weigh the logs, especially if they were small in size. Then they devised a system of buying all the logs by weighing the truck loaded. I found that fewer small and partly rotten logs were left in the woods under this system, and that the net pay for the logs gained about four percent. That added to my profit.

Under Dale Jones's supervision, the Jones Logging Company designed the first collapsible tower for logging. They mounted it on a large ex-army tank carrier truck. The tower had four steel posts and was about 60 feet high. It was devised so that it folded in the center. Now it was possible to make smaller landings and to take advantage of the lay of the land for easier yarding. Less time was lost in rigging up and taking down the rigging from a spar tree. This idea of a collapsible logging tower was soon improved on until all the modern logging machines use a telescoping tower with hydraulic power. These improved yarders are mounted on wheels and self-powered from one landing to the next. The move that with the old style yarder would take several days can now be made in just a four or five hours. No longer is a high-climber needed to top the spar tree and put up the rigging. This modern yarder carries the fully rigged spar pole from landing to landing.

Byron Ames, singing logger

Byron Ames worked for the Willamette Valley Logging Company's camp for eight years as a donkey puncher. He usually operated the machine that did the loading of the logs onto railroad cars in the days of steam power. When the steam motors were replaced by diesel powered heel-boom loaders, Byron Ames had trouble adjusting to the improved method of loading logs.

Byron Ames, top. The other loggers are unidentified. Ames operated the steam-powered machine that loaded logs onto rail cars at Windling, Oregon. Photo courtesy of Charles V. Ames.

Byron Ames had an unusual singing voice and a natural musical talent. He sang on all public occasions. On Sundays he liked to sit on his front porch and sing. People driving by would stop their cars and stay to listen. When he stayed in the bunkhouse at camp, in the evenings most of the crew would gather after supper just to hear him sing. He loved his work in the logging camps and though often urged to try out as a public entertainer and singer, he never felt like giving up the work he loved loading logs. It is regrettable that Byron Ames lived before television and video tapes.

His son Bill was much like Byron in singing, making music and telling stories. Some of Bill's music and stories were put on video tapes. It would take the writing of a book to do justice to Byron Ames and his remarkable family. He and his wife Eda were parents of ten daughters and five sons: Janice, Jenny, Nancy, Geraldine, Jimmy, John "Jack", Bessie, Hazel, Mary, Dorothy, Delilah "Pops", Lloyd, Charley, Bill, and Laura "Toots."

All five of Byron Ames's sons went into the logging work. Lloyd, like Charley, was a skilled high-climber, limbing and topping trees to be used as spar trees for yarding in logs to cold decks or landings.

The author bucking a log in a clear-cut area with a modern, lighter-weight power saw in 1958. Note axe behind him, and measuring tape hanging from his belt. No more measuring with a stick or by striding along a log top. Photo from the author's collection.

The Gypo Logger

When a thrifty workman could save enough money to make the down payment on a machine, he could start his venture in logging. Quite a number of men who were just workmen on logging crews made good in their logging ventures. The stumpage could be bought at a low price. There was lots of timber in the Tillamook Burn selling for one dollar a thousand board feet, and worth at the dump about $30 a thousand. With wages what they were at that time, money could be made if you did not have to build many miles of road, or had a dirt road you could use in the dry weather of summer.

The first time I got into the logging game, I went into partnership with a man who was a good rigging man for one of the companies I had worked for. They gave us a contract to log a tract of timber leased to us for one dollar each, a large donkey engine to yard the logs and a Caterpillar tractor to make roads with. These machines had seen much service. The tractor broke down the first week we used it and we had to make the repairs and pay for the new parts. Then my partner would meet with the equipment salesmen in the beer parlor and buy equipment that we needed but could not afford. He, no doubt, was thinking of the days when he worked for an established logging company. Our timber was up on a mountain and there was not enough of it to warrant a rock road. At the end of summer I sold out my share in the venture and got away from the worry of paying the bills for equipment that we could get along without.

The first Gypo contract we had, the stumpage for large old-growth fir and spruce was at six dollars a thousand, and we should have made money on the deal.

In my years of logging, on two occasions I tried Gypo logging. The first venture failed when my partner overspent. The other venture failed when we spent our money building the road into a unit of what proved rotten timber. This was before the cull logs went to the chipper. I expect that at a later date some other Gypo logger used our road and salvaged the cull logs we had to leave in the woods.

Oney McDonald & Bart McKinney

These two men worked in the logging camps for several years, Oney at the falling and bucking and Bart on the rigging. In 1942 they managed to get a small tract of timber to log. As it was winter time, and no other work available, they fell and bucked this timber. Then when good weather came, they managed to rent a tractor and log the tract. They had only a small distance to haul the logs to the river, where they could be floated to the mill. Then they contracted another tract to log and haul the logs to market. They had somehow managed the down payment on a tractor. This timber was mostly old-growth spruce near Manzanita Beach, near Nehalem, Oregon. They had no money to pay for the falling and bucking of this timber. It was late winter, and very little work available. Most of the logging by the larger companies was at an elevation where snow was too deep for working.

They asked the two Palmer brothers, who lived nearby, to come and fall the trees and they asked me to buck the logs and us to wait until they got their money after marketing the logs. Since there was no other work to be found, we agreed to cut their logs for them. The unemployment insurance at that time paid only $14.00 a week. They would pay us $6.00 a day when they got their money. Well, it seemed we would be better paid at the end of two months than to draw on the unemployment insurance.

Just when the Palmer brothers had finished the falling, the Markham & Callow Company started logging on Sweet Home Creek, a branch of the North Fork of the Nehalem River. They were free, but I still had about ten days work to finish cutting McDonald and McKinney's trees into log lengths. They were fortunate to have a dry April. That let them get their logs to market before the rains started in again in mid-May and lasted until after the Fourth of July. So, in due time, we got our pay.

McDonald and McKinney did not stay in partnership long. Oney was a sharp dealer. They went their separate ways and did quite well in their logging ventures. When Bart McKinney made enough money to buy a ranch in Central Oregon, he quit the logging work and left the woods.

I mentioned about Oney's sharp deals. When it came time to settle up, I had marked an extra day in my time book. When we went to Bart's house, he came out and took my time book and put in two extra days. When Oney looked at the time, he said, "I know you spent at least one day hunting," so I let him scratch off a day. Now we were all ready to settle. Bart sure knew his partner, and I got two extra days pay for waiting for my money.

It would be difficult for a small so-called gypo logger to get into the logging business under today's rules and regulations at today's wages, price of stumpage, cost of insurance, and with regular pay periods. Also, there are no longer the small tracts of timber available. We will have to wait for a lot of new growth. That will not be ready for logging until the loggers of the next generation are old enough to go to work.

Elmer Storm

Elmer Storm came to Oregon in the late 1940s. Like a lot of people who came west during and just after the Depression, he soon made good. These people were good workmen, and ambitious. Most of them arrived here in Western Oregon with little more than the clothes on their backs, driving an old beat-up car that they somehow managed to make get them here.

Elmer went to work in a logging camp. Somehow he managed to get the use of a tractor and a small patch of timber to log. In a few short years, he had by hard work and good management made a success of his logging business.

The Windfall Bucker

Crew sent with only axes and a two-man crosscut saw to buck and clear away a big windfall, blown down across a road. From left: Henry Peck, David Stone and Elton Yates.
Photo courtesy of Elton Yates

In the days before the power saw, with its its smoke and noise, I liked to work at bucking windfalls. The windfall bucker went into the forest ahead of the tree fallers and cut into logs all the downed trees that were sound enough for milling. Any cull trees also had cuts made in them. In the hot days of summer, the windfall bucker could work in the shade. The man bucking the felled trees into log lengths usually had to work out on a hot side hill or down in a canyon with the sun's heat reflected off the felled trees. The windfall bucker was by himself. He did not have to worry lest he be endangered by the hazard of falling trees or falling logs. There were squirrels to chatter to him and birds singing. At lunch time he could have the birds we called "camp robbers" for company. These birds were always looking for the chance to snitch a bite. I have had them try to take a sandwich out of my hand as I paused between bites. Then there were deer seemingly coming to check on us and wonder why we had invaded their domain.

One time I had another man working nearby, another windfall bucker. We each had our territory marked out. He had the habit of, if a tree looked like it was easy to cut, slipping over into my area to cut up the tree.

One day as I moved my tools up to a large tree, I heard a noise on the other side of the tree. Just as I put my hands up on the tree and prepared to make good my claim, here came a woolly black head. Then with a loud woof, and our heads only inches apart, the head jerked back. I don't know which of us was the most startled. That big old black bear had quicker reflexes than I, for he was gone crashing through the brush before I could get into action.

There are some men who prefer to buck windfalls. They come to camp asking for the windfall bucking work. This story is told at Foss. One wintertime, a middle-aged man came and asked for a job bucking windfalls. The bullbuck, who is the man in charge of the cutting crew, took the man on up the railroad tracks to where the next unit would be cut. He pointed out the hillside to be made ready for cutting. He told the man to go to the bottom and first cut out the windfalls in the canyon.

Thinking this new workman was an old, experienced log cutter, he never bothered to go down in this deep canyon to check on the man's work. Each day as he walked by on the railroad he could hear the noise made by the pounding of the steel maul against the steel wedges, the usual sounds by which a windfall bucker can be located.

After a few days, the foreman thought that the bucker's noise should have moved from the bottom of the canyon. When he went to check on the bucker's work, he found the man had a fire going in front of a bark shelter, and that he had sawed well down into the first cut, then inserted the two steel wedges. He watched to see this man get up from his comfortable seat, go over and strike the maul a few strokes, then go back and sit by the fire and take out his pipe and resume smoking.

Elmer Whitmore

When I started in grade school, Elmer was one of the older boys in school. He was three years older than I. I always looked up to him. He was one of those seemingly slow of action, but whatever he did, he did well. He was exceptionally good with a baseball bat or shooting a .22 rifle. At an early age he was able to do a man's work, but he would rather hunt than work. Like all we men who grew up on the Chehalem Mountain farms, he learned to use an axe and pull a falling saw to cut down trees. It seemed only natural that he eventually went to work for the big timber cutting companies.

Elmer Whitmore always seemed to be able to get along well with his fellow workmen and the men in charge of the work. Elmer was rated as one of the best at cutting those large trees. He could always see the best place to lay the trees to save the most saw logs. He seemed to be able to find work at any logging operation. He worked mostly in the Coast Range all the way from Clatsop county in northwest Oregon south into northern California.

Soon after he married my sister, Esther, he inherited a small farm on Chehalem Mountain. After he chose to devote his working time to cutting the large trees in our forest lands, he sold the farm. He bought two houses, one in Garibaldi and the other at Twin Rocks on the ocean. He lived first in the Garibaldi house, later in the Twin Rocks house. When his wife's parents retired from farming, he moved out and had them move into the Twin Rocks house.

When the State of Oregon took over the vast Tillamook Burn area, it was the practice to salvage only the best wood in the fire-killed fir trees. To get the land ready for replanting, it was necessary to cut these dead snags. Some areas were contracted out for cutting by skilled workmen. A camp for housing convict labor was established near the head of Wilson River, just south of where the paved highway crosses the summit of the Coast Range mountains. Young men were given the chance to earn some money and learn to be skilled workmen while serving out their prison sentences.

This worked out very well. Elmer Whitmore was hired to oversee some of their work. He trained these men to cut the dead trees with a power saw and to build fire trails and primitive roads all through the Tillamook Burn area. Elmer worked from 1954 to 1964 training prisoners to become useful citizens of our state. Declining health and reaching retirement age caused him to leave this work that he found very rewarding.

Elmer had one bad habit that eventually caused his death. He liked to drink good whiskey and smoke cigarettes. On a weekend he could spend the entire two days just sitting in a chair at home smoking and drinking. Finally his doctor told him either to quit smoking and drinking or die. We were all surprised that Elmer on his retirement cleanly quit both the smoking and the drinking. But he had only a short time to enjoy his retirement. His doctor had told him to do no heavy lifting or strenuous work. Elmer loved clams, and in Tillamook Bay there were lots of clams.

It was early spring weather when Elmer, with two companions, decided to go dig some clams. The sand bars where large blue clams lived in Tillamook Bay could only be reached with a boat. Elmer and his companions decided to go for the cockle clams in the sand along the main channel of the bay, near where the Miami River enters the bay. There was one problem. For centuries this stream had been dropping its mud into the bay. To reach the cockle clam bed you have to cross a mud flat in which your feet sink a good ten inches with each step. Cockle clams lay in the sand,just barely covered. Sometimes a bit of the closed shell is visible. The best way to uncover them is with an ordinary garden rake.

Now these three men struggled through the mud flat and raked out their limit of clams, but Elmer never made it back to his car. That struggle through deep mud was too much for his weakened heart. He collapsed right there in the mud. It was more than two men could do to extract him from that mud flat. Other help soon came to their assistance, some by boat, others across the mud flat. I don't remember how they managed. If they used the rake handles, it would have taken six men.

I wrote this account on April 7, 1996, thirty-one years after the death of Elmer Whitmore.

George Willey

George Willey was a man with many talents. He ended his working days logging in our forest lands. He spent more years falling trees and cutting them into logs than at any other trade. George at a young age learned blacksmithing. He became skilled at welding and tempering metals, a skill he often found useful in later days. His working with metals led him to study geology and mining. He became a self-taught mining expert. This mining was underground work. He soon decided he would much better like working with an open sky over his head. He became interested in the collecting of agates and a skilled lapidary.

In order to live and support his wife and son, he went back to blacksmithing and decided to locate at the little sawmill and logging town of Gilchrist, on the east side of the Cascade Mountains. It seems that while living at this town in the pine forest, George did whatever work that came his way. He did some carpentry and house-moving. He also cut some pine trees into logs for the mill.

It was probably in 1936 or 1937 that George Willey came to work in the Oakridge area, at the time the highway up Salt Creek was improved and paved. That quarter mile of roadbed just a short distance above the second bridge was giving the workers a problem. The road right-of-way crosses an earth fault. It seemed to swallow load after load of base rock. Fortunately, a rock ridge just below the bridge furnished an abundance of material. Since George was experienced in blasting rock, he was given the work of blasting out that rock. They found the best result was gained by dumping in pieces of rock as large as they could load into the dump trucks. It took days of work to fill this space of roadway. Each morning, the workmen would find that the previous days load of rock had sunk, and several more loads of rock must be dumped. Even today, after 60 years, that short section of roadbed is still sinking. Every year, after winter is past and drier days are here, that stretch of roadbed has to be repaired. Another layer or more of paving is added to bring the roadway up to the required standard.

The large fir trees here on the west side of the Cascade Mountains were a challenge to George Willey. He just had to learn to fall and cut the large trees on the North Fork of the Willamette River into logs for the mill at West Fir.

It was in 1947 that the Pope and Talbot[1] sawmill at Oakridge was built. George Willey started to work on Pope and Talbot's cutting crew, up the main branch of the Willamette River. By then, the Willey family was well established in their house at Oakridge and taking an active part in the life of the community.

In May of 1961, I was working for Pope and Talbot. George and I soon became good friends since we both worked on that cutting crew and both enjoyed hunting, fishing, and working with agates. We started a companion-ship that was to last through the next fifteen years.

Pope and Talbot was and is a company that appreciates good workmen and safe working conditions. They found that the safest way to cut their timber was for two men to work at the felling of the tree and the cutting of logs. The work was paid on a day basis, and the company furnished the tools and transportation from the mill right to our work site. Some of the trees to be cut were at a

---

[1]Pope and Talbot no longer logs in Oregon at this time. At this writing (June, 1997), their main logging operations are in British Columbia, with a small amount of work being done in the Black Hills of South Dakota. Editor

high elevation where deep snow made it impossible to work in the winter time.

This meant that in the winter only a crew of the older men could be sure of work. Each spring more men were needed on the cutting crew. This meant that in the summer months we older and more experienced men were given younger, less experienced, men as partners.

George Willey and I were both looking and working for the day of our retirement. Since I was nearly a year older than George, my retirement day came a few months sooner than his. I really enjoyed the five years that I worked for Pope and Talbot. All I had to do was get on the bus, ride out to the work place, and give a good day's work for my wages. My nights and days off were mine to enjoy.

It so happened that George and I were cutting partners the last few months that I worked. We each knew the work well. Our

George Willey, photographed in 1968 at the small forge where he manufactured the "Willey B" rockhammer he and the author invented.
Photo by the author

foreman was a man very sparing with words of praise, but after George and I had retired, he was known to say, "Those two old men sure made the work of falling cutting the trees look easy. The best thing they could do was to take a young partner and show him how to get the work done!"

George and his wife were great for going places. They visited every place of interest in all ten of our Western states, especially where agates were to be found. George had a well-equipped shop for cutting and polishing agates. He was very generous in helping others in cutting and polishing. He loved to take people out on agate hunting trips. He seemed to always know where to look, and usually he found the largest and best agates.

I remember one time we were chukar hunting in the eastern part of our state. A chukar, or Persian partridge, is a grayish brown bird with red legs and bill. They are great birds for laughing at you. They'll try to scare you by jumping up right at your feet, then they'll get out of gunshot range and laugh at you.

Always in our game hunting we carried a few tools in our packs, just in case we should find agates. One day in the Stinking River Basin, I saw George kneeling down in the sage brush. When I came to see his find, he had discovered a really large agate and had out his rock hammer. He was just about to strike. I cried out, "George! Don't break it!"

"It is too large for me to carry," he said.

It was a large piece of white plume agate. Rather than see him break it into smaller pieces, and destroy this rare agate, I was willing to try to carry it the mile and a half to our pickup. He would carry my gun and it was all downhill. It was quite a hard pack to carry that 75-pound chunk of rock, but I made it. Then at home, on my 20-inch rock saw, I turned and cut this agate into two pieces and gave George one half of it. My half will be made into a valuable set of bookends.

The best agates are often embedded in hard rock. George and I found there was a need for a better tool for breaking the rock to free the agates. We finally designed a better tool. We named it "The Willey B Rock Hammer." We found a steel made in Sweden that would stand the hard

Logs descend in a flume from a landing sited at the top of a steep hill to a mill at the bottom.
Building a road on such steep ground would have been impractical. Photo by Asahel Curtis,
courtesy of Washington State Historical Society.

battering and not dull nor break. But George Willey was the only person who could get the right temper to this tool.

The Hampton Butte area was a great place for agate hunting. Much good agate ore was just lying on top of the ground, but people—since it was public ground, with no restrictions—came and hauled it away by the truck load. On one trip to that area, George wanted to show me a whole agatized tree, but when we came to the place, there were only small chips left. We saw the tire tracks of a large truck. George told me of another place, on private land, where the property owner used his tractor to cover up a whole agatized tree to save it from being broken up and hauled away, regardless of the posted signs.

One time George and I came to one of these posted ranches just at the beginning of grain harvest. In a wheat field near the road, we found two men with a combine and they were in trouble. A drive shaft had broken. They were wondering where to get a new shaft. George stopped and talked to the men. He asked, "Have you got a forge at your ranch?"

They said, "Yes, a small forge." They said they had used it when they had horses for power.

George told them he could weld that broken shaft. Needless to say, the rancher was happy with the welding, and George had their permission to agate hunt on their ranch. We were fortunate to find this small forge, so that George could use it in tempering our tools for working in hard-rock digging of agates.

George and Mary Willey spent a lot of time going to agate shows and rock pow-wows, displaying his agates and selling his polished agates and the jewelry that he made but his love was for finding and helping others find and dig out agates.

One year in early spring, they went to Arizona to help some friends put on an agate show. He spent several days taking people out to find fire agates. The weather was cold and windy. George, on the way home, became ill and had to go to the hospital. He never recovered.

About that same time, I found that I no longer could make those hard trips to dig agates. We had to give up on making the Willey B Rock Hammer and stocking up for my lapidary work. I could find no one to temper steel like George.

Bim Spalanger, Best Bullbuck

Bim Spalanger had charge of the cutting crew of the Pope and Talbot Logging Company at Oakridge, Oregon. It was on the second day of May, 1962, that I went to work cutting trees for this logging operation. I rate Bim Spalanger the best bullbuck of all the men I worked under in my many years in the logging camps.

When in April of 1962 I found myself in need of work, I contacted Fred Hayes, the business agent for the labor union that I had held membership in for twenty years. I told him I needed a job. I wanted to work in a union-contracted operation. Our labor union had just completed an agreement that gave members a good retirement plan. Fred contacted the Pope and Talbot bullbuck. They immediately phoned me, offering me work. They had changed their timber cutting from paying the workmen by the thousand board feet to day work. I had for years worked at getting paid on the rate of how many thousand feet of logs I could cut in a day's work. This day-work was just what I wanted for the few years I need to work before my retirement. No more having to spend hours figuring log rule, or the many other extra work hours needed to keep an operation going. For the previous fifteen years I had had the responsibility of managing the cutting crews in contract logging.

Bim Spalanger was a foreman who was a good judge of the abilities of the men under his directing. He always first looked over the timber to be cut, then saw to it that the men started cutting in the best manner to save the trees. He found that the best and safest plan was for the men to work together and fell and buck the trees to be cut. One man operated the power saw to fell the trees. The other stood by to put in wedges if needed and to see that the proper amount of holding wood was left on the far side from the operator. When the tree was on the ground, one man marked the logs while the other cut the tree into saw log lengths with the power saw. This way, one man had the chance to see any danger, and perhaps avoid an accident. Many times this plan has saved a man from injury.

Spalanger made good use of my ability to control the falling of trees and either put me on the roughest ground or close to the road. The only time my trees were crossed up or one hit a stump, it would be caused by rot in the tree. If there is stump rot in the base of a tree, it may not fall in the place where we intend to lay it. I only recall one time that my tree hit a stump, and that had to happen in a place where all who passed by could see it. I had to endure a lot of good-natured kidding. Some would say, "Look, Old Man Brunson missed a lay." Spalanger appreciated a workman who did a good day's work, but he had no patience for a shirker.

As usual, I was soon chosen by the cutting crew to represent them in any meeting with the company representatives for negotiation where there was a difference in the working agreement. I must say this about the Pope and Talbot company management: They were the most courteous to the crew's representatives of any company that I ever worked with. I'd say it's an unusual

Dirk Rutger, age 80, came to Menefee's logging show on Lost Creek in northern Tillamook County, Oregon in 1942. It was during World War II, when all the young loggers had gone to war. Rutger wanted to do his part. After he cut this one big tree he said, "This work is too much for me," and turned in his tools. Menefee hired him then as watchman. The ancient logger brought his trailer house to the site and worked four more years.

Photo from the author's collection

happening when the representatives of a company are courteous to the workmen.

At Oakridge both the company and the union negotiators realized we had a good system, and both sides were wanting it to work, realizing that both company and men must work together if the company is to prosper and keep in operation.

Spalanger was not much for paying compliments. After I had retired he said where it would get back to me, "That Old Man Brunson was a good man to have on a crew of log cutters. He could do any work needed, but the best thing he did was to teach those young men how to get the work done!"

Some of the other crew men thought I regularly did more than needed for a good day's work. I did note that I mainly was given an inexperienced working partner. Now usually it takes two or three years for a young workman to learn to be a good tree faller. The last year I worked, Spalanger put Jim Owen to work with me. When I retired, he gave Jim the head faller's responsibility. We even got mentioned in the wood worker's magazine, the old man in his last year and the new man in his first year in the logging woods, cutting trees.

Due to the agreement, I worked more than a year past the retirement age in order to get into the new retirement benefits.

I was surprised on my last day of work when I got off the crummy at the company office. Spalanger the bullbuck, the logging superintendent, and the general superintendent were all there to wish me a "Happy Retirement" time. I don't know how Spalanger got his nickname. In the local phone directory his nickname was included. He was a good manager, and eventually was given the top job in the logging operation.

CHAPTER FOUR

OF BIG TREES, FOREST FIRES,
CAULK SHOES, AND TIN PANTS

Logging the Luckiamute River Country

When I started working in the logging camps in 1923, the only way into the Luckiamute
River area was by railroads that were built into the Coast Range by logging operators. This was
before good roads and log trucks, radio and television.

Long before the ax men with their hand saws, the greatest destroyers of our forests were
uncontrolled fires. After such disasters, nature in its own time reseeds and grows back our
forests but nature may reseed it scattered or too close for best regrowing and may leave the tops
of ridges bare. When a tree is cut we can count the rings, one for each year it has lived and
grown. We can tell the good years for growing by the spacing between the rings and its age by
the number of rings. I have found fire scars healed over as much as 250 rings inside of a tree.
After clear-cutting our steep mountainsides, we get the trees to grow faster by hand replanting.

One of the best stands of old-growth trees in our State was that section of our Coast Range
mountains that lies between the Yaquina River and the Salmon Rivers. The Little and Big
Luckiamute Rivers flow to the east from this area and the Siletz River flows to the west.

I spent most of my working years at logging this area, mostly cutting in from the western
side. The best stand of old-growth trees I cut was on Gravel Creek, a tributary of the Siletz
River. These trees were from 400 to 600 years old and all sound trees. We logged six million
board feet off 40 acres. Most trees that age have reached maturity and have started to deteriorate
due to disease or other natural causes. We were fortunate that these old trees were still healthy.

Today most of our forest lands are in tree farms. Our forest lands and national forests are
carefully managed. The steep mountainsides must be clear-cut and then replanted. Each little tree
must be carefully placed in the ground. Where the lay of the land permits, selective logging can
be practiced and all mature trees salvaged. All logs that can go into lumber or plywood are sent
to the proper mill to get the most from them. Everything else is made into wood chips, so there is
no need of burning to clear the land for replanting. It seems that the trees can best be cut at age
60 to 70 years. We can no longer wait the 300 years until a tree starts maturing or is past its
fastest growing years.

It gives us a lot of pleasure to drive through the timber in our forest lands over well-kept
logging roads and see the new growth of trees. We never gave a thought when we were cutting
those large old trees that we would see the day that these mountain sides would again grow trees
large enough for the mills. Today so much of our timber land has been set aside for parks or for
other reasons that we are having to cut small trees at their fastest growing age. Many of those
old mature trees we are forbidden to cut are rotting faster than they are growing. I speak from
experience. I spent near 50 years working in our forests.

Charley Ames topping a spar tree in the 1920s.
Photo from C. Ames collection.

In Search of Western Oregon's Big Trees

The Bureau of Land Management has left a section of forest land in its natural state. We think it is well to save a stand of these old-growth trees so that future generations may see them. This timberland is on the north fork of the Siletz River near the Polk and Lincoln County lines.

My grandson, Greg Wolf, and I decided to go with a video camera and take some pictures. We had to wait until after the fire season, until the roads throughout these forest lands were open to everyone. We stopped in Fall City for Charley Ames, as our guide.

Charley Ames was a good high climber and rig-up man when we first worked in the same logging camp, back in 1936, the year the Social Security started. We were working for the Balderee Logging Company at Foss, Oregon. Balderee was the smallest of three companies logging on the Big Luckiamute. The other two, the Spalding Company and Willamette Valley Lumber Company, had movable camps. When they finished logging an area, they just hitched their shops and houses onto a train engine and pulled the buildings up closer to standing timber in the next area to be logged off. The loggers never had very far to travel to their work.

Charley had lived, worked and hunted all through these mountains. Charley showed us where the house stood when the Ames family lived at the Black Rock logging camp—Byron and Eda Ames and their fifteen children, including Charley. He pointed out the place where the one room school stood, where he and many of his brothers and sisters went to school from 1914 through 1920. Now only trees are growing in this wide place in the little valley. Black Rock camp was really a company-owned town with all the buildings needed in a large logging operation. The Willamette Valley Lumber Company had first built the railroad from Dallas to Fall City, then up the Little Luckiamute River. At the Black Rock camp they had a depot and several miles of side tracks on each side of the main line. The company railroad was the only way the logging families could get to Dallas, or any place out of Black Rock.

Western Logging Company had their main camp near the big pond at Valsetz. Cobb and Mitchell owned all Valsetz until Boise Cascade took over all the standing timber that was left. Boise Cascade had a plywood mill at Valsetz in the 1940s.

When the Southern Pacific bought the railroad it became a common carrier and the CKS

The school at Black Rock in the Luckiamute River area, photographed in 1915. The recently logged-off hillside behind the school has since grown two more crops of trees. By 1996, a third crop of trees had grown to 14 feet tall. Photo from collection of Charles V. Ames.

Holding and the Western Lumber Company came in and set up their own logging operations. The Spalding Company's holdings were to the southwest and the Western Timber Company's lands to the Northwest. In those days, all logs were hauled over railroads. Southern Pacific
added a motorized coach to carry passengers from Black Rock to Dallas. It made two runs a day. This railroad coach was the only transportation for upwards of 1,000 people living at Black Rock and its satellite lumber camps.

Charley Ames at times worked as a brakeman on these railroads. The three major logging operations up the Little Luckiamute required many miles of rails, often on steep grades. This required specially built locomotives with names like Shay, Hystler, Climax, or Saddleback. The Saddleback locomotive had the water tank draped over the boiler to give added traction. On three sides of the camp were the tracks for turning around the locomotives, sets of tracks called 'wyes'. With the coming of diesel and gas propelled motors, these steam-powered motors were phased out of the logging camps. Then, with the coming of tractors and log trucks, the railroads were taken up. In the mid-1940s, the post office at Black Rock was closed and all the buildings torn down. Now the railroad right of way and the town site are over-grown by young trees.

From Black Rock we drove up a well-kept road in a northwesterly direction. We took a left-hand road up near the summit of the mountain. Now it was all downhill on crooked and rough roads. We finally came in sight of the old trees. We crossed a primitive bridge and more

Charley Ames, showing his climbing tools and technique in 1996 at the age of 82. In the 1950's he headed logging operations for Vulcan Lumber Co., owned by himself and his partner, C. L. Fallert. They also owned South Coast Lumber mill, shipping lumber from Crescent City, California. Photo by Mark Gulbrekson

rough roads. Some of the roads were badly overgrown with young alder. A new growth of trees screened off the old trees.

Charley had brought his climbing tools and we had planned on pictures of him climbing one of those large old trees. But we found no tree that we could get close to. We noted survey stakes at the bridge across the stream. Plans were under way to build a new bridge and improve the road so bus loads of people can get in and see these trees.

Driving along the South Fork of the Siletz River, we stopped and looked at the place where that stream had been dammed to make the millpond at Valsetz. We saw only young alder trees growing where once were our favorite fishing waters. There were trees growing where the mill stood, and where a town of more than 1,000 people lived at one time. There was no sign of the railroad that Cobb and Mitchell built from Monmouth and up the Big Luckiamute River in the 1920s. Before the railroad was built, all the logs were floated down the river. We went over the ridge to the north fork of the Siletz River to see the section of old-growth fir trees on Board of Land Management property. We found the place grown up with new growth, mostly alder. We were not able to get good pictures of those large trees 400 and more years old because of all the new growth around them.

Charley knew of a large old-growth fir tree near Valsetz We took the road that goes South and east and over to the Big Luckiamute River. Then the first right hand turn, and there right along the road was a monster fir tree. We measured this tree at shoulder height, and it was more than 11 feet in diameter at the height where it would be cut by the chain saw, and well over 30 feet around. The safety rope on his high climber belt was only 28 feet long. There was no way that we could get video pictures of Charley climbing this tree. We did take some video pictures with Charley and this large old tree, holding his climbing tools. This old tree has rot killing it from the top and another type of rot in its base. In time more of the top will blow

An inclined logging railroad. The slope is even steeper than it looks. Rail cars were hauled up and let down on cables. A loaded car going down provided a counterweight causing the empty car to go up. They passed midway on the double track section near the top.
Photo courtesy of Washington State Historical Society.

out and eventually the stump rot will weaken the root system. Then a strong winter storm will uproot this majestic tree. I would judge this tree to be more than 600 years old at the present time.

There is still a lot of good wood in the center of the tree. It has a 34-foot and a 43-foot log that can be salvaged. This means at least 10,000 board feet of sound wood, worth several thousand dollars, if delivered to the mill. This tree is an example of the way trees were left in the 1920's. It is what we loggers call a stump rotter. It is good that this tree has been left so that anyone can see it.

We came home by the Valsetz-Fall City road, which was built in the late 1920's, after many of the logs had been hauled out by railroads. Where this road crossed the summit, Charley showed us where the Spalding Company had logging camps and an inclined railroad to take the loaded cars of logs down the mountain to Black Rock. An incline is a steep railroad that goes straight down a mountain. The log cars are let down by cable controlled by a specially built steam engine called a "snubber." Usually an inclined railroad has two tracks up the mountain. As a loaded car goes down, it counter-weights an empty car, causing the empty car to go up. By building an inclined railroad, the company avoids having to build many miles of switch-back tracks.

Charley Ames only went two years in High School. Soon after his 14th birthday, in the fall of his sophomore year, after an illness that put him so far behind his class he thought he'd never catch up, he told his dad, "Find me a job and I'll go to work."

He started as a whistle punk at the Black Rock logging operation, in the days of the steam logging machines. A wire attached to the whistle was taken out into the area to be logged so that the whistle operators would be near the men setting the cables onto the logs. By jerking down on the wire, with the right tension on it, the whistle would blow to give the donkey puncher the signals. When the gas or diesel powered motors and the radio operated signals came, the whistle punk's job was phased out, and Charley went on to higher work.

The Luckiamute Today
Written March 31, 1997

We have some pictures of the Luckiamute area. Some of this land has been logged three times, re-planted yet again, and now bears a new growth of trees. When it was first logged, back when I started cutting in our forest lands in 1923, we cut only the best trees. Any tree with any sign of rot, or with limbs growing too close to the ground to give us a marketable log, we left to reseed the area. We often left doubtful logs, or logs hard to get out. About 1950 and later, with the invention of power saws and tractors, and with portable yarder towers to replace the use of spar trees, all the trees could be used. They either became lumber or, with the chipper, were turned into raw material for pressed wood. Now we can go in and salvage all the timber that was left behind, even the slashings that were formerly burned.

Where forests were naturally reseeded, timber frequently grew too close together for the best growth. Lack of sunlight contributed to stump rot. When we hand-plant young nursery-grown trees, we can properly space them so each can have room to grow. Planting young nursery-grown seedlings also enables us to get many more crops of trees from the same land, since these young trees have at least two years head start over seeds planted by nature. Where nature reseeds, forest lands require 70 years to produce a crop of grown trees. Hand-planting and new forest management methods let us harvest the same area again in 40 years.

In 1994 we took video pictures of the forested Boise Cascade Company logging road. Three years later, in the spring of 1997, we went to take some still pictures to illustrate new growth of timber for this book. We were surprised that the area had already been clear-cut. Trees too small for logs went into wood chips. The land had been already made ready for replanting with new, young trees, properly spaced. By now, young trees are growing there, producing a new forest for our next generations. This is making good use of our forest lands.

The Tillamook Burn Fires of 1933 and 1939

It was in August of 1933 that the first fire in the area called the Tillamook Burn, aided by a strong east wind, swept from the east side of the coast range mountains, a distance of near 40 miles, and over an area about that wide, in just three days.

This fast-traveling fire left large areas untouched. We were fortunate that a moist current of air came from the ocean and stopped the fire from going any farther, and brought a soaking rain to put it out.

We had two other major fires at intervals of 6 years. These later fires burned out of control until most of those old-growth trees were also killed. It is a heart breaking scene to have to helplessly stand and watch a fire eat its way into a hillside of large old-growth trees and in a matter of minutes go sweeping up a mountain side, changing green trees to blackened snags. All vegetation is burned. Not a thing is left for the wild life to feed upon. When winter comes most of the wild animals that escaped death by fire will die of starvation.

Tongues of flame climb tree trunks, feeding on dry moss on the bark. On reaching foliage, the fire "crowns". Burning material picked up by wind scatters, starting fires in new areas. Photo dated October 14, 1952. Courtesy of Oregon Historical Society, Neg. #63476. Oregon Journal Collection

Some of the privately owned timber in the Tillamook Burn was logged. It was up to the State of Oregon to take over and supervise the salvaging of this fire-kill area nearly as large as some states. It took about 40 years to clear the hill sides of dead snags and plant young trees. Now, over 60 years after the first disastrous fire, the whole of the burned area is green again with new trees, and the first logs from "the new Tillamook Forest" are going to the mills to help solve our log shortage. Roads have been built along all the ridges, and are kept open, so that this forest will never be so devastated by fire again.

The forest fire that devastated vast areas of Tillamook County's forest ranges in 1933 darkened skies for three days. Billions of soggy, charred fir needles that had blown out over the Pacific Ocean washed ashore. They lined the county's beaches along the high tide line in brownish black piles, almost knee-deep. When the 1939 fire started, my older brother, Wayne, and two of my younger brothers were part of a crew that had spent two months cutting trees on a Tillamook County hillside, up on the mountain in the Wilson River country. They were working for a small logging outfit near Jordan Creek. The logging road angled up the mountain and through a cut on a small ridge to reach the landing.

They had rigged the spar pole and they had moved the yarding machine in and had started to yard up and load the trucks with the newly cut logs. All work stopped when this rapidly moving fire seemed sure to reach their cuttings.

The logging foreman thought it a good idea to move the yarder and other equipment into the near-by road cut. They had just built a new sled for the yarder. For a large yarding machine, two logs each about four feet in diameter and near 50 feet in length were framed together. They should last for near the machine's lifetime. The tractor pulled the yarder into this nearby cut. The men hoped that it would be safe there from the fire. They bulldozed up enough dirt to close both ends. To close off the top, They used some sheet metal strips that had been brought in for building an office and shop. They laid these metal sheets across the top to completely close in the yarder and other equipment that the fire could damage.

They had just managed to get this roof in place when, with a roar and intense heat, the fire drove them off the mountain. Every crew member jumped into the crummy or pickups. They raced that fire down to the highway on the Wilson River.

When the foreman and a few of his crew felt it safe, they went up there to check on the fire damage. It was a desolate sight—everything burnt black. To their surprise, the carefully laid sheet metal was nowhere in sight. No doubt the terrific heat and wind had crumpled up those metal strips and blown them into a deep canyon. The yarding machine lay there in the ashes of those two large logs and the cross-timbers and bolts that had held them together. The bolts were so twisted they could never be used again. Most of the cables were also so badly burned that they would be unsafe to use in logging.

I was working at that time on the cutting crew at the logging operation of the Markham and Callow Logging Company. We were cutting in the basin of a small stream between the Little Nehalem River and the main Nehalem River. For two days we could see smoke billowing up from the main fire. The logged-off area had well-bulldozed fire trails, but the large area of standing green old-growth fir trees was unprotected. The rigging crews kept on working to get in as many of the cut logs as they possibly could before the fire got to our area.

We of the cutting crew were sent out to build fire trails. We worked on the ridge that divided the drainage between the two rivers, hoping to stop the fire and save the stand of virgin timber. We were fortunate that by then we had a light west wind from the ocean. The day the fire reached our lines, we got a light drizzle. That stopped the fire from crowning and slowed it down. We still had the problem of spot fires jumping our fire trails and a slow creeping fire along the ground.

Our outermost fire trail had still nearly two miles to go to reach the main Nehalem River. The last mile ran down a steep mountainside with a thick brush cover. We had already worked three days in smoke and heat. The light drizzle of rain only made the smoke more irritating to breathe.

It was late in the afternoon when we neared the river. We knew that it was there, and with a shout of relief the first man broke through the brush and stood on the river bank. He shouted "River!"

In turn, each man on the crew shouted, "River!"

Trees burning like candles in Tillamook County, Oregon, in 1949. Even spacing suggests that trees burning in the foreground were hand-planted after the 1918 fire, thirty years earlier. Probably the hill in the background with closer, less evenly spaced trees, was naturally reseeded. Fires like this kill both man's and nature's attempts to grow green forests for future generations.
Photo courtesy of Oregon Historical Society, Neg. #49536

It was a tired and dirty crew that stood on that river bank. Now we could lay down our tools and rest. Our crew boss told us to just leave our tools on the river bank. We did not have to climb back up that mountain and follow the fire trail a good five miles back to the end of our logging road. The Nehalem River was so low we could wade across. On the other side was a good trail, and it was only a mile downriver to where, on a primitive road, a crummy waited to haul us back to the main camp on the highway. Someone asked about our tools on the river bank and was told, "Leave them." Men patrolling the fire trail would take them back up the mountain to the logging road.

When we got back to headquarters camp, it was near our regular quitting time and we could go home. But not all of us were free of that fire trail. Some of us were asked if we would come back and help patrol the fire trail until all fire danger was over. Crews of men would have to patrol both day and night. I was fortunate to be able to go home and get a good night's sleep before my turn.

One incident enlivened the crew one night. Each crew had about a mile to patrol. One man could stay and keep a fire going while other men patrolled each way. It always seems that sounds

Driving through the Tillamook Burn on the Wilson River Highway in the winter of 1949-50. This area has now grown a new crop of trees, shading highway and stream and blanketing the hills with living green timber. The car is a 1948 Keizer. Photo from the author's collection

in the night are more frightening than those in daytime. I can imagine no more hair-raising sound to hear at 2:30 a.m. on a dark night than the scream of a cougar. This is what happened on the first night after the completion of the fire trail. The men out on patrol were so startled that they all came running back to the fire nearest to the logging road. When the relief crews arrived in the morning they found all six of the night patrol around the one fire.

In all my time in our forests, I have only seen two cougars, and only fleeting glimpses at that. I have found their well-marked trails and the places where they stay for rest on the territory they each claim for their own. I envy those men on patrol that night. Hearing the cougar scream was the thrill of a lifetime.

Editors note: According to the U. S. Forest Council, we are enjoying new growth of timber 30% faster than we are cutting at this time. Yet, because current laws, and the efforts of well-intentioned, but often-ill-informed, activists so restrict logging, the United States must now import lumber from Canada, and from Central and South American countries where sound timber management and logging practices are either unknown or not practiced. For lack of logs, hundreds of lumber mills all over our nation have had to close, throwing their workers out of their jobs, and being forced to cut their losses by selling their

The same scene of the Wilson River Highway, now a drive through the new Tillamook State Forest. Reseeded and hand-planted young trees have grown to again cover these coast hills with living green. It has taken almost half a century to repair forest fire devastation of 1933, 1939 and 1949. Photo in the summer of 1997, by the author and his son, Paul Brunson.

Measuring Logs

When I first started bucking logs at the Manary Logging Camp near Yachats, Oregon, the logs were cut in 16- to 64-foot lengths. We used an 8-foot measuring stick. In falling the leaning spruce trees, often long strips were pulled out of the butt of the tree. When we needed a measuring stick, we would go and cut one of these long slivers off the stump and trim it down to suit us. At this camp we usually allowed from one to two feet of trim. We did not have to be very careful as long as the trim was over one foot.

At some camps they wanted at least eight inches and not more than twelve inches of trim on the logs. When I worked on the cutting crew in the pine timber south of Bend, this was the rule. In the pine timber they just gave us one saw to fall and buck the trees. That saw was too limber for use by one man, so after limbing and measuring a tree, the two men cut the logs, one

---

machinery to operators in these other countries. This is not good for the earth as a whole. Also, the timberlands where harvesting is forbidden become prey to wildfires that sweep vast areas clear of all life. The trees, bushes, and animals living in these fire-ruined areas all perish. Because of the intense heat forest fires can generate, even the bacteria that help make the soil good for growing plants can be killed. Clay soil can be fired hard so no seed can take root. M. W.

Measuring a log containing 7,000 board feet of lumber to see if it's legal size for trucking, at Menefee Logging Co. in 1955. Photo from the author's collection.

man on each end of the saw. I didn't like this way of bucking logs. I bought a used 7-foot regular bucking saw. Since I was best at this kind of bucking, my partner did the marking and measuring out of the 32-foot logs. He made a cut mark with his ax where the cut was to be made.

I noted that often he was hitting the end of his measuring pole. I had cautioned him lest he shorten the measuring pole. In due time the check-scaler came to see if we were getting the 8- to 12-inch trim on the logs. He found that our logs were short of the 8-inch trim and went back and took two feet of scale off all the logs we had cut for the past three weeks.

n the early 1930's, we started using 50-foot steel tapes to measure logs. We found that a horseshoe nail, bent around the loop in the end of the tape, worked best to hold the tape's end. We could stick the sharp point into the wood and it would hold the end of the tape until we measured out the log length. Then we'd give a sharp jerk and the end would come loose so we could wind up the tape. Then a spring was devised so that the tape would self-wind. This saved the hand-winding time. Now we snapped this self-winding tape to our belt, fastened the end, and with our saw in hand hurried to the next cut. When we didn't have to be exact on the trim length, we just made a scuff mark with our caulk shoes, jerked the tape loose and let it self-wind. In the days before self-winding tape, we often could just step off the log length. I found that my step always came right on the mark at 36 feet. Often, after marking the cut, I would check with the tape to make sure that my stepping was on the mark and my measure accurate.

The Logger's Shoe

On steep hills the strongly built, leather high-top shoe with steel caulks in the soles was a very important item for the men working in the logging camps. This was especially true for workers in the burned timber after the bark had come off the trees.

When we felled trees with a springboard we needed caulked boots in moving the spring-board. This board had a half-circle shaped metal bit that fitted into a notch we chopped into the tree. With the proper footwork, we could swing the board as we needed to position it as a place to stand while we chopped the undercut and sawed the back of the tree. The wood at the base of a tree is timber-bound. The saw used to stick in it. Now with our power saws we cut the tree at near ground level. The modern power saw will cut that cross-grained wood.

It is only on the steeper ground that steel caulked shoes are still used. On ground that can be logged with a tractor, a composition nonskid shoe sole can be used. These are made of more pliable leather than the steel caulk shoe requires.

The problem was getting a shoe that fitted your feet. There were always several small companies making logger boots in our Western towns. Many of these shoemakers would make shoes to the customer's foot measurement. Some of the ready-made shoes were a trial to the wearer. The longer you wore them, the harder and stiffer the leather became.

Quite often, when a logger bought new caulk shoes, he threw his old well-worn shoes under the bunkhouse. In a well-run logging operation no man could go to work without logger's shoes on his feet. If we came to camp looking for work with no money or credit to buy a new pair of caulked shoes from the commissary, we could just look under the bunkhouse and find a discarded pair. These old shoes would be hard on a man's feet, but he usually could make do until he had credit at the commissary.

I could wear the West Coast shoe. This shoe was made in Portland in 1920, but the plant was enlarged and moved to Scappoose, Oregon. In February of 1923 I went to the West Coast Shoe Company's plant and bought a pair of 16-inch-top loggers that cost $16. At that time my wages were 50 cents per hour. At that wage, it took several days to pay for the shoes. Today that kind of shoe is priced at near $200, but the logger's pay has increased so it takes fewer days work to pay for his shoes. I note that men working on ground not so steep are wearing a non-skid shoe with composition sole. That is cheaper than the shoe with steel caulks in hard leather soles.

For more than thirty years I had my shoes made by John Johnson. He had his shop in the basement of his house in Portland, out on Foster Road near 70th Street. Johnson came from Norway as a young man. He had learned shoemaking before coming to America. Like many of his fellow countrymen, he heard of the good wages paid in the logging camps. Soon he learned of the uncomfortable shoes his fellow workmen were wearing. He thought he could make a more comfortable shoe for them. He carefully took the measurements of several feet, went back to town and made up shoes to the measure of each foot. The men were so pleased with these shoes that they told Johnson he was a better shoemaker than a logger. When I first met the man, he had come up to the camp on the Kilchis River delivering shoes. The next time I needed shoes, I went to his basement shop. He took the measurement of my feet, made and sent me the shoes. He carefully filed away each man's foot measurements so that we had only to send him the money and he would build our shoes and send them to us.

While writing about logger's shoes, I will write on an experience we had at the logging operation run by Joe Flora at Foss, Oregon.

On the cutting crew, the bullbuck also did the marking and scaling of the trees. Logs that were good for making plywood were in demand. Care was taken to cut these logs in peeler lengths. In our coast mountains, often small streams will come out of a basin that is circular in shape. Such a basin is near ideal for logging. There was a little flat, the little canyons and ridges all leading down to the center of the circle. Here the little streams all got together and ran down a steep canyon to Cook Creek and the railroad. It was about 3,000 feet from the railroad track up to where the yarder, spar pole and landing would be. It was about 1,500 feet to the top of the ridges. For most of this unit, the trees would be felled down the steep hillsides.

In measuring trees on steep ground, it is best to always move with your feet pointing uphill. You get better traction with the caulks. Walker, the scaler, one day made the mistake in measuring off the butt log of a tree by just walking down the tree. It could be that this tree in falling and sliding away down the mountain had shed its bark. This made a slicker surface than a tree with bark. Walker lost his footing. He slid down the tree and broke a leg.

It was late afternoon. Here was a man with a broken leg and the nearest stretcher for carrying him was a rough three quarters of a mile away at 2,000 feet lower elevation. It happened that one man on the eight-man cutting crew knew some first aid. He took charge. He sent one man for the stretcher. He got Walker into a more comfortable position, straightened his leg, got the leg back in place, cut some wooden splints, bound the injury with strips from his shirt, and soon had the man ready for the long hard trip down the mountain. For a stretcher, he had us cut two poles. He turned two raincoats inside out, put the poles through the armholes, then buttoned the coats. We placed Walker on this makeshift device. We tied him with leather shoe laces so that he could not slip forward on the steeper slopes. At times, it took all six of us to get him past some of the rougher spots.

By the time we got down to the landing, help with a regular stretcher met us. The 3,000 feet on down to the railroad track and the speeder to take us in to camp was on a good trail with steps cut in some of the steeper places.

We were 'til after quitting time getting the injured man to the office at camp. I remember we laid him, stretcher and all, up onto the counter. Neither at camp nor in the nearest town at Wheeler was there an ambulance. The owner of the butcher shop in Wheeler had a panel delivery truck. We finally got him to come and haul the patient to the hospital at Astoria. At the hospital the doctor was so impressed by our care. He found the broken bones in place so that he had only to put on a regular cast.

Tin pants, hickory shirts, and hard hats

All the logging camps kept in their commissaries a full line of loggers' clothing. We wore wool underclothing, and waist-length pants. We cut off the legs to show at least six inches of boot. As the trousers were waist length, not bib overalls, we wore a pair of heavy-weight suspenders, too. For winter, on our rainy West Coast, we had pants and coat of heavy weight canvas. We waterproofed then by coating then with paraffin wax. We called them 'tin pants' since they were so stiff that when you put them on the floor waist-down, they would stand alone, with the legs straight up, especially after they had received a coat of fir pitch at our work.

Shirts were of sturdy cloth called 'hickory', or of wool. We usually wore a hat made either of soft felt or of the same material as the tin pants.

It wasn't until 1950 that the hard hat was designed. We of the cutting crew were the first to wear hard hats. I can remember several men who in my days of felling trees were saved by their hard hats from severe head injuries, and I was one of them. Many of the rigging men would not wear a hard hat until it was made a safety requirement. One logging crew made fun of us on the cutting crew in our hard hats. Then one of them was slapped on his head by a line and carried out to the ambulance on a stretcher. We on the cutting crew all smiled when the next morning each member of that logging crew got out of the crummy wearing a new hard hat.

Danger Trees and Widow Makers

When asked about injuries to forest workers, I reply that I have received or seen most of them. I spent most of my time on the cutting crews. At it's best, cutting trees is very hazardous work. Cutting fire-killed trees like those dead snags in the old Tillamook Burn was especially dangerous. We cut only the fir trees. Hemlock no longer had market value. This was before the development of a market for wood chips.

In this tree-cutting work, we find a wide variety of widow-makers. When falling the fir trees among the dead and rotten snags often one of them is knocked over and comes right back toward the stump. One good safety measure is to always get away from the stump when the tree starts to fall. Often a piece of limb will break and come down at a very high speed. Sometimes, when the tree we have just cut knocks down one of those dead snags, it will come right back to where the faller is standing for safety. In such cases, I would have to watch to see on which side of my standing tree the falling one would fall, and then step to the other side of my tree and safety.

Clear-cutting is always the safest. You can fall your tree away from the standing trees. Selective cutting is always dangerous as you have to fall your trees into other trees left standing and uncut.

I remember one time, when I had to fell some small trees into other small trees, a small piece of limb came back and hit me. It cut a gash on my forehead that took three stitches to close up, and knocked me out. I never saw nor heard that piece of limb. The first I knew, I was flat on my back beside my powersaw and it was still running.

The most severe injury I had was when a green tree that my felled tree hit was thus caused to uproot and throw its top at me. I survived only because my nephew, Walter Whitmore, was working nearby and heard me yell and came to my rescue.

The most spectacular accident was when a dead spruce tree uprooted and a root threw me up into a small fir tree. It held me there until others came and cut the root so that I could be helped back to earth.

Often workmen receive injuries by doing the work the way their foreman directs. We were clearing a landing for the yarder one time. Our foreman told us not to cut the smaller fir trees. The tractor would just push them over with the blade. This was easier to do than to dig out the stump after we cut these trees.

Now in falling a dangerous tree, two rules of safety should be followed: 1. Do not try to save the power saw; give first thought to your own safety. 2. Make sure of a safe escape route.

I was falling a big, dead, leaning spruce tree, and the safe escape was blocked with a clump of small trees about 18 inches in diameter at ground level. I cut the undercut with the seven-horsepower saw with the 70-inch bar. I made the perfect initial cut, and cut out the undercut clean. Now these two saw cuts, and taking out that wedge of wood, unbalanced this leaning tree. It started to uproot.

I had cut a lot of leaning trees but never before had one uproot after just making the undercut. Usually, if the tree uproots, it is while you are cutting the back and just as the tree starts to fall. I lost a split second when I threw the saw clear so that the tree would not fall on it. Now I was in real trouble. A spruce tree has large long roots near the surface of the ground. One of these roots grabbed me and flung me into that clump of small fir trees. My chest hit a fir tree. There I was with my left leg pinned to the tree and the wind bashed out of me.

I must have let out a yell when I hit that fir tree. My partner and the cat skinner quickly came. With a much smaller power saw, they cut the tree I was fastened to, but that didn't help. That fir was held by the other fir trees in the clump. Here I was twenty feet up in the air and yelling, "Be careful, or my leg will break!"

Now the men had to dig away the dirt and cut off the spruce root that held my left leg to the fir tree. It was only six inches thick where it held my leg, but a good two feet in thickness where they had to cut it.

It seemed to me that it took those two men a long time to cut me free and get me back to earth. I was fortunate not to have a broken leg, instead of only a badly bruised leg. I did have a broken left rib. My falling partner got me into the pick-up and took me into town to the doctor's office. He estimated that it would take three weeks at the least before I could go back to using a power saw.

That was good news to my wife and me. We had planned to make a trip east to visit relatives in Kansas on our vacation that year. With an automatic clutch in our new Chevrolet station wagon, and a good bandage around my ribs, there was no reason to put off the trip.

Our daughter Shirley and her husband and their new baby, David, were just ready to leave on their vacation to show his parents in Minnesota their first grandchild. We made the trip together. Had I known it would take four weeks for my broken rib to heal enough for me to go back to cutting trees with a power saw, I would have taken time to detour into Arizona to see the Grand Canyon on our way home.

Log Chutes and Flumes

Before the tractor and other modern road-building equipment were designed, log chutes, flumes, and railroad inclines brought the logs down steep mountain sides. I remember the first log chute I ever saw in use. It crossed the Kings Grade road on Chehalem Mountain, near the steepest part of the road. The land was owned by Tom Burns. It was nearly all covered with second-growth fir trees. This chute was made to carry the trees harvested above the road down to the mill. The chute had to be steep so the logs would slide down it. It had to be fairly straight or a fast-moving log would jump out of the chute.

Travel on this mountain road was with horses, as few automobiles of that day in 1912 would try to make it up this steep grade. To send logs down the chute, the driver of a team of horses would drag a number of logs down to the landing at the top of the chute. Then he would send his helper down to the public road. They had a cow bell rigged so he could ring it by pulling a wire stretched from the road back up to the landing. When a team of horses came by on the road, the helper rang the bell signaling to stop sending logs down the chute until the team of horses was safely past. The high bank on the uphill side of the road made plenty of room for a loaded wagon and team to pass under the chute.

These logs were cut into 16-foot lengths. They started a log on its way down the chute by using horses. As the horses got a log started sliding down the steep slope, a hook in the end of it would come loose. At the bottom of the chute, the logs landed on the ground. It was better when a pond was at the bottom of a chute so the logs could land in water.

Long winding flume brought boards from a rough-cutting mill in the hills to a planing mill on the Columbia River for Bridal Veil Timber Company, where they were finished.
Photo courtesy of Oregon Historical Society, Neg. #7872.

In the summer of 1925, I worked for Frank Nelson on Baker Creek, west of McMinnville, Oregon. He built a chute about 2,000 feet long to bring logs down a mountainside. They yarded the logs into the upper end of this chute by a steam donkey. They first landed the logs on the ground, until the mill pond was completed.

At the head of tidewater of the Miami River, near Garibaldi, a log chute was also used. It looked like a vertical gash on the hillside. We could watch from the highway, and see the logs come down the chute and splash into the water. Often it has proved more practical to build a water trough called a flume in which to float small logs, or boards cut at a mill up in the hills, to a millpond or to a finishing mill at a lower level on a road or a railroad.

Logging Machinery
Written March 28, 1997

Improved machinery has changed and replaced all the older machines and way of cutting trees and getting the logs to market.

When the oxen and horses were replaced in the woods by steam engines to pull the logs to the loading sites, the engines were called "donkeys." The first donkeys were fueled with four-foot lengths of wood. Men had to cut the wood, and one man had to put the wood into the fire-box to heat water in the boiler to make the steam.

Later, to save manpower, donkey engines were designed to burn crude oil. An operator called the "donkey puncher" could control the amount of fuel fed to the fire to generate steam. There was no more need for men to split wood, nor for a fireman. It still took nearly an hour to get up steam. A man went out and built the fire before the loading crew came to work so as to have a head of steam up when the men arrived. Somebody still had to cut the wood for the fire to heat the crude oil. Then when gas and diesel fueled engines were designed, this early fireman's job was eliminated.

In Barney Klahn's story of logging machinery, in Chapter Five, he mentions the donkey sled. The donkey was mounted on a sled made of four-foot-thick logs, a sled strong enough to stand the strain of moving such a large and powerful machine.

When I worked on the cutting crew at the Markam and Callow camp on the Nehalem River in 1949, I helped build a donkey sled. They needed two men who were good with axe and saw. We on the cutting crew were not paid day wages. My partner and I were paid at the rate of our log-cutting average, which would be more than the day wage.

It took a lot of sawing, ax and adz work to build these sleds. The foreman carefully marked every cut and every hole to be bored. We used two logs at least 50 feet in length. First, we beveled the ends so the sled could slip over small logs and other bumps encountered in moving the heavy, powerful machine. These two logs were placed at the proper distance from each other and tied together with cross-pieces fastened in with iron bolts about two inches in diameter. We mortised all cross members into the logs. All the nuts were counter sunk. The crosspieces were close enough together that extra equipment could be carried on the donkey sled when being moved. Our sled was so solidly put together that a cable attached to one runner could move the entire machine and it would last through many years of stress and hard usage. The engine frame was mortised into the sled and bolted down. The fair lead on the front end was likewise fastened. The sled carried three drums for spooling the three cables used in moving the donkey. The main line was the largest. It was more than 1,500 feet long.

The haul back line had to be more than twice the length of the main line but smaller in diameter. The small line called the "straw line" was the longest line. It was a utility line of many uses. It had to be pulled out by hand all the way then hooked to the haul back line to pull out that line. These lines all run through blocks. The main line's block was largest. The haulback line and the pullout line ran through smaller blocks called side blocks. These blocks were sometimes pulled out to their positions with the straw line wrapped around a stump called a "siwash". Some of these blocks were a hard one-man carry up a mountainside.

CHAPTER FIVE

YARDERS AND SPAR TREES

Don't Let the Donkey Run Away

In my Whitney Logging story, in Chapter One, I tell of a yarding machine getting away as they were moving it from one side of the landing to the other. When yarding logs with a spar pole, we can log only one-half of the landing. Then, to pull in the rest of the logs, we must move the donkey, that is, the yarding machine, to the other side of the spar pole. Now, when moving a yarder on steep hillsides, you must at all times use a safety line. This is a cable attached to the donkey sled and to a stump uphill.

I have helped move yarder machines up mountains steeper than a house roof. On one move, we used two block hitches on both the main line and the haul back line to lift this heavy machine. With a double two-block power, we could have taken our yarder straight up. But we also used the safety line whenever we had to change holds. We took three turns around a stump uphill from the donkey and spiked it with six railroad spikes, driven in pairs, one on each side of the cable. The shape of the heads of these spikes, with a flange on one side, made them ideal for holding cables in place.

On this occasion at Whitney Logging Camp, they didn't spike the holding cable. The foreman set three men to holding the cable. Of course, their weight and strength could not hold the heavy machine from sliding downhill and being lost.

In the old days when we could only move logs to the mills by railroad, we had to use a skyline up to 3,000 feet long to bring logs to the loading site. When I worked for the Manary Logging Company near Yachats in 1923, it took a crew three weeks to move a donkey engine across a canyon and up to the yarding landing, a distance of about 3,000 feet, then rig up the spar pole and put up the skyline. It took a lot of work to clear a way for it up that mountain side through the felled and bucked trees. Many of those big logs made single-log loads. I once found three trees where I could only make the cut at 80 feet. I asked the loader if he ever got logs longer than 64 feet. "Yes," he said. "It would help if we could get a carload of them."

I knew he would get a carload of 80-foot logs from those three threes. Each log would have 3,000 to 4,000 board feet of lumber in it. Most of the railroad cars on this private railroad would carry 12,000 board feet of logs. In medium-size logs, 80-feet was the length desired.

Wood-burning steam yarding donkey and crew of Bear Creek Logging Co at Swenson, Oregon. The donkey engine's stack ends with a bulbous screen to keep sparks from escaping and starting a forest fire. The donkey puncher stands under a shed roof, with his hand on a lever. Two others are choker setters. At left, ax in hand, is the wood chopper, probably the youngest member of this crew. Photo by Clark Kinsey, Neg.#50, courtesy of University of Washington, Allen Library, Special Collections.

Horace M. "Barney" Klahn worked on rigging crews during the first half of this century. In 1962, he wrote this account of moving the heavy machines that powered the pulling of logs from where they were cut to the cold decks or landings from which they could be loaded onto rail cars or trucks. Charley Ames made Barney Klahn's writings available to us. They are published, lightly edited, with Mr. Klahn's permission.

MOVING YARDERS CROSS COUNTRY

by Barney Klahn

Before we moved logs from the woods on trucks, railroads were the main means of transportation. Where timber was too far from the railroad loading sides, and railroad spurs could not be built to the timber, those logs had to be yarded into cold decks, then yarded or "swung" from there to the loading sites. To do this, yarders called cold deck machines, not unlike a regular yarder, had to move themselves out to the spar tree where the cold deck was to be built. They had to pull themselves cross country, using their own lines and rigging.

We transported the yarder by rail or truck to as close as possible to the cold deck tree and unloaded it. To make it pull itself to its setting we ran a strawline and small block out in the direction of the pull. Sometimes it was only a couple of hundred feet. Other times, it might be even the length of the main line, depending on the lay of the ground. We fastened the block to, or close to, a stump or tree we were pulling to, and threaded the eye of the strawline through the block then took it back to the yarder. This line was threaded through the main line fair-lead and fastened to the eye of the main line. (Another way to do this is to put the eye of the strawline through the main line fairlead and fasten it to the main line, then taking the bite of the strawline out, put it in the block, previously hung as explained.)

Now we could pull the main line out through the fair-lead with the strawline, and fasten the main line to the stump or tree, called a 'tail hold'. The yarder was now ready to make its pull. The boss, or hooker, generally stood on the head block on the front of the yarder, balancing himself against the fair lead. From there he could see what was going on and signal to the yarder puncher.

This sort of hookup moved the yarder on fairly level ground. If it strayed off its course, with the haulback line, you could pull the nose of the sled sideways. To do this you pull the haulback line out through its fair lead to the side of the nose of the yarder in the direction you want to steer the yarder to, and tail hold the haulback line to a stump or tree.

To move the yarder over steep ground, you needed more power. A large "moving block" was taken out to the tail hold you were pulling to and fastened with a heavy strap made of probably 1 1/4-inch diameter cable. You did this with the strawline, and generally one or two men followed along with the moving block in case it hung up in some manner. They had to free it, as the strawline wasn't strong enough for pulling the block out of a hang-up. Then, with the strawline, you pulled the main line out; threaded it through the moving block, and then pulled the eye of the main line back to the yarder and fastened it to a lug or eye located on the outside of each sled runner, close to the front of the runner. This way, the sled could be steered one way or another by moving the eye of the main line to lugs on either side of the sled. You could also steer the sled by using the haulback line.

To move a cold deck machine cross country, we generally needed several pulls of different lengths, depending on the lay of the ground. This involved a lot of strawline pulling and block packing, but it was the only way to get the machine to where it was needed. These yarders had fairly large runners, as they had to go over the top of any downed timber. If the sled runners straddled a small tree, they just pushed it over and went over the top of it. Brush, they generally didn't even worry about.

When the Port Angeles Western Railroad was being built towards Forks, Washington, on the Olympic Peninsula, around 1930, after the right-of-way timber was felled and bucked, they used a machine called a Clyde, probably after the manufacturer. They yarded the logs off the right-of-way with this small 3-drum yarder by using an A-frame fastened to the nose of the sled. The A-frame was made of two small logs about 1 1/2 feet in diameter and about 60 feet long. Each log, or leg of the A-frame, was bolted to the nose of a runner and held up at about a 60-degree angle by two cables or guy lines fastened to the top of the A-frame .They ran back and anchored to the rear of the sled runners, one line to each runner.

The Clyde moved itself down the railroad grade just as a cold deck machine would do. Every several hundred feet, it would be turned sideways to the railroad grade, with the top of the A-frame jutting out over the side of the grade. A couple of guy lines hung from the top of the A-frame were run out to and fastened to stumps opposite the pull of the yarding line. The machine then could yard all of the right-of-way logs off the grade and into small cold decks along the side of the grade. There they were out of the way of the crew laying tracks. It was a slow process, but it got the job done.

Once on the upper Olympic Peninsula, in 1940 or 1941, we were moving a cold deck machine over timber that had been felled and bucked. It had been a thick stand of trees. The logs just about covered the ground. Everything went fine as we moved over the tops of the logs, but then we came to a grassy slope. The timber ended abruptly where this field began. The slope ran down towards a small creek. Across the creek was a dirt bank about 10 feet high. The tail hold for the pull was on a big stump on top of the bank. The grass was very wet, almost swampy-like.

The donkey puncher wanted the hooker to run the haulback line underneath the yarder and out the back end and choke it to a stump. This would be a brake in case the yarder took off when it hit the wet grass. The hooker didn't think that necessary, so we kept going. When the sled dropped onto the wet grass, the slope of the field was steep enough that the yarder really took off. It gained enough momentum that, when it hit the bank across the creek, the sled's nose buried itself so deep it covered the head block. The mud really flew!

We got the main line unhooked from the stump we were pulling to, and wound it up onto the drum. I, being the youngest man in the crew, had to crawl under the sled between the runners, in all that mud and water, taking the strawline with me, and fasten it to the eye of the main line. Then we could pull it out the back end and fasten the main line to a stump, so the yarder could pull itself out of its predicament

Bloedel-Donovan had a cold deck machine working out of their Pysht River camp near Clallam Bay, Washington. It was powered with a big gasoline motor. We built a cold deck across the Pysht River canyon from the railroad grade. If memory serves me, it was about two miles upriver from their camp.

One morning we had to take on fuel. To get the gasoline barrels to the yarder, bringing them across the canyon from the railroad flat car that brought them, we ran a haulback line through a lead block near the yarder, across the canyon, to choke it to a stump on the hillside above the railroad car.

The haulback line went over the top of the fuel tank at the rear of the sled, and also over the middle of the railroad car. This haulback line thus acted as a skyline. We put a small rider block on the haulback, with a weight attached to the yoke of the block to keep the yoke down so that the wheel would ride on the haulback. We hooked a couple of chains to the yoke to choke the barrels with. The tail hold being somewhat higher than the yarder, the barrels would descend by gravity to the yarder. To pull the chains back to the flatcar, we ran a strawline across the canyon, through a block, and hooked it to the rider block. This set-up worked pretty well, and we transported 33 barrels of gasoline, two at a time, and filled the yarder's fuel tank.

When we finished that cold deck, we had to move the yarder down to the river, across it, then up a steep hill and across the railroad tracks. Everything went fine until we got across the river and started up the hill. It was so steep, the carburetor float wouldn't let the gas come through. The Hooker made me get up onto the side of the motor. He gave me a small can of gas to pour down the throat of the carburetor.

"You better not let that motor die!" he told me.

Tail tree rigging crew in 1935 at Valley Lumber Co., now called Willamette Industries. From right: Charles V. Ames, his brother Lloyd "Bud" Ames, Dan O'Dell, W. Robinson.   Note straw line coils in foreground.
Photo courtesy of Charles V. Ames.

I didn't really want to ride that machine up the canyon wall, but I was more scared of the Hooker than the ride. Everything went fine, though. We snorted on up the hill and across the railroad tracks to our next setting.

The longest single hold-pull I ever was on was at the Ozette Timber Company on the Olympic Peninsula, southeast of Ozette Lake. It was flat country and the timber was mostly cedar, hemlock and spruce. We went out the full length of the main line, 1200 feet or so. We had to do very little side-pulling to steer the sled. It was so easy we could have almost took a nap.

One gypo logging outfit working out of Sappho, Washington, about 1940, had to move their yarder up a steep hill and build a cold deck on top of the hill to yard logs down to the railroad line. I wasn't working for them when they moved the yarder up the hill, but I helped build the cold deck, then move the yarder down the hill and across the railroad tracks. There we rigged a small spar tree and yarded the cold deck logs down to the tracks where it could be loaded out. On the downhill move, if my recollections are correct, the main line was run under and out the back end of the yarder to be used as a brake to keep it from running way. The haulback line was used as the pulling line. You sure don't want a machine getting away from you on a downhill run.

Yarder engineer P. Blackburn, hoses off a wood-burning donkey. In front are hook tender Floyd Blackburn, left, with wood cutter Charley Ames, center, age 19. The log behind the young men is actually one runner of the donkey's sled. Photo courtesy of Charles V. Ames.

One old-time logger told of moving a steam yarder across a river where they knew the water was going to get high enough in the firebox to put the fire out, but there would still be some space between the water and the bottom of the boiler. They pulled the yarder into the river and made sure they had a full head of steam when the water put the fire out. They'd cut some planks ahead of time to put into the firebox where they would float. They poured kerosene onto the planks and set it afire. It kept up enough steam to get them across the river. It was maybe a crazy idea, but it worked, and that's what counts. Sometimes it takes a little horse sense and ingenuity to get the job done.

A lot of things were done years ago that they don't do now. Some of then, you wouldn't dare to do. Crescent Logging had a track-side tree rigged about a half-mile west of where the present US Forest Station is now at Shuwah, about five miles east of Forks. The logging railroad ran parallel to Highway 101, and less than a hundred feet from it. The top and buckle guy lines were strung to stumps across the highway. Traffic went under them without any bother.

I imagine that during the 1930s many a yarder was drug across a highway or country road without anyone thinking twice about it. Now almost everything pertaining to yarding and loading logs is mounted on tracks or rubber tires. The track machines are moved on lowboys, and most of the rubber-mounted machines move themselves. Quite a change.

Spar tree that has been brought to a site carved from a hill and rigged
with lines for raising it upright. Lines go to stumps high on a hill off left.
                    Photo courtesy of Horace M. Klahn

When no suitable tall tree grew on the site chosen for a landing, in the days before portable steel towers, loggers had to bring a spar tree to the site and raise it. Horace "Barney "Klahn's account has been slightly edited.

RAISING A SPAR TREE

by Barney Klahn

In logging by the high-lead method west of the Cascade Mts. in Washington, Oregon, and Northern California, before the introduction of steel towers, when there wasn't a standing tree where it was needed, a spar tree had to be raised. . . .

The first thing was to fall a suitable tree as close to the landing as possible and cut the top out. A 24-inch diameter top was about as small as you wanted for safety sake and a little larger was better. Most trees were at least 100 feet long and it didn't hurt to have one somewhat longer to get more lift when yarding. The tree was then dragged to the landing site and the butt placed into position where the (spar) tree was to stand. If no Cat was available, you used the yarder or rig-up machine and its lines. Generally a pad was made for the tree to stand on so that it wouldn't settle into the ground with usage, which would slacken the guy lines. The pad was several small logs eight to ten feet long, lashed together with a length of strawline. The pad was placed under the butt of the spar tree and pulled under the tree butt a couple of feet so that when the tree was raised it more or less centered on the pad.

The top of the tree was raised off the ground two or three feet and the bark peeled off where the top and buckle guys were to be hung. With a Cat, this would be a simple job. If not, a chunk was pulled under the top, raising it a foot or so with a line, or holes were dug under the tree at those points to get the guy lines and straps around the tree. Lines were shackled to a stump near the top of the tree and then you took the bit of that line around the butt of the tree, laying it in notches cut in on each side of the butt, so that it wouldn't drop to the ground. Then you took the other end of the guy line out to a stump towards the top of the tree, tightened it up and spiked it, using a couple of wraps.

Next we pulled the tree under the top, raising it a foot or so with a line. Sometimes we dug holes under the tree at points where we needed to put the guy lines and straps around it The tree had to be tied so that it wouldn't slip end-to off the pad when it was being raise .We shackled one of the guy lines to a stump near the top of the tree and then took the bit of that line around the butt of the tree, laying it in notches cut in on each side of the butt, so that it wouldn't drop down to the ground.

By tightening the top By line and slacking the lower line, the spar tree has been lifted at an angle.

The steep slope at left and in the next photo shows how a flat landing was carved from a steep hillside.

Logs dangling from skylines will be pulled from down the hill to the landing by the spar tree.

Spar tree half-raised, butt end resting on pad, and guyed by wires to hillside stump.   Photo courtesy of Horace M. Klahn.

Two men notched guy line stumps—5 to 6 stumps for top guys and 3 to 4 for buckle guys. It was their job to see that the stumps were notched properly. The Hooker picked the stumps ahead of time so they knew which ones to notch.

The first things to be hung on the tree while it was on the ground were the plates for the top-guys. Tree plates were made of steel about three feet long, 1 inch thick and about four inches wide. Each plate had two hooks, made of about the same material, and welded to the plate, one near the bottom and the other about the middle. These hooks kept the guy lines and block straps from slipping down. The plates were spiked to the tree with railroad spikes. Four plates were hung at the top, spaced evenly around the tree. These were to keep the guy lines and block straps from cutting into the tree thus weakening it.

Next, three guy lines were hung around the tree at the top and fastened with sleeve shackles. Most of the time the ground around a landing was hilly enough so that a stump on a hillside, located high enough to get sufficient lift, could be used as a tail hold to raise the tree with. Two of the guy lines were strung out to stumps opposite the direction that the tree was to be raised. To string guy lines to their stumps, a small block was hung on a log or stump several feet behind the guy line stump. The eye of the strawline was pulled out to that block and threaded through it, then the eye pulled back to the tree and hooked into the eye of the guy line for that stump. The guy line could then be pulled to and around it's stump. The two back guy lines were spaced far enough apart, in a "V", so that they could keep the tree from swaying sideways as it was being raised.

The third guy line was hung so that it could be fastened to a stump located in the direction the tree was being raised, along with the other two guy lines for holding the tree up after it was in place. Then all unused rigging could be taken down from the top of the tree.

The two guy lines on the back side of the tree were taken out to their respective stumps. One wrap of line went around its stump, securely in the notch. That was to be held by the haulback line. The other line was placed around its stump with two wraps. Railroad spikes were driven into the stump to hold the wraps in place. These spikes were left just a little loose so the line could slip around the stump as the tree raised.

Three men stayed at this stump while the tree was being raised, two men to handle the guy line and the third to tap the wraps with a hammer so that the guy line would slip.

In raising a spar tree with a sled-mounted machine, the sled had to be tied to a stump so that it wouldn't move forward in a pull. Lead blocks for the main line and haulback line, were hung somewhere in front of the sled to keep it in place. Trying to tighten a guy line with the main line running through the fair lead, if the pull is off to one side, would only pull the sled in that direction.

Left: The same spar tree, nearly vertical, guyed to stay upright. High lead lines will bring logs from downslope on each side of its ridge.

Photo courtesy of Horace M. Klahn.

A spar tree, fully rigged and yarding logs, with some of the bark still on it. The two dark smoke clouds behind the tree are from donkey engines, a yarding donkey, left, and a loading donkey nearer the spar tree. The small white puff is from a train engine. A log is being loaded onto the second of two flat cars. Telescoping metal spar towers on wheels have replaced such spar trees, and logs travel now on trucks.

Photo courtesy of Washington State Historical Society.

We hung a lead block on the hillside for raising the spar tree. Another block was hung at the top or near the top of the spar tree with a strap to be used as L-block purchase for raising the tree. The strawline was threaded through these blocks in such a way that we pulled the main line out with the straw-line, then threaded it through both blocks. The end of the main line would be shackled to the same stump the lead block was on, or one nearby. This was the tail hold for the L-block purchase. Enough railroad spikes were carried to each guy line stump to safely secure the line after the tree was raised. We hung a pass block at the very top of the spar tree. We threaded the pass line or pass rope through this block, leaving both ends of the line on the ground. Later, after the tree was raised, that line would take the rigging man and the remaining rigging up the tree.

We raised the tree with a three drum hoist. When the tree was ready, the haulback line was taken out through a lead block in front of the machine, and then out and shackled to the eye of the guy line on the back side of the tree, the line with the one wrap around its stump. Three men on the other stump gave it slack. From then on the hooker did most of the signaling.

Tractor pushing a large log to a landing at Menefee's camp on Lost Creek, 1942. Loading donkey is under shed roof at left. Logging road comes in from the top of the photo. Note the headgear. Hard hats to prevent head injuries were not invented until 1950. Photo from the author's collection.

Raising a tree is a fairly slow process. The two guy lines on the back side of the tree have to be rendered around their stumps as the tree comes up, also to keep the tree from swinging sideways and falling to the ground. The third guy line would be left slack at the foot of the tree and tightened after the tree was raised and in place.

Just before the tree was virtually in place, the hooker, by judging the "belly" in the guy line that the three men on the stumps were handling, would have them spike the two wraps of their line tight to the stump. Generally they used about 16 to 20 spikes on the first wrap and 4 to 6 on the second wrap. This would hold the tree until they could put the third wrap on later with the haulback line. Top guy lines with no more than 1 1/2 to 2 ft. of belly arc are pretty tight lines.

The haulback line holding the other guy line was held tight. As the tree was raised into place, it was slipped just enough to leave the proper belly in that line. The men then spiked the first wrap and proceeded to put the other two wraps around the stump and spike them tight. To do this, they hung a small block near the guy line stump, preferably behind it, then pulled the strawline out to that block. They threaded it through the block and back to the guy line. They fastened it to the guy line with a rigging chain in such a spot that slack could be pulled on the guy line to put wraps on the stump. The strawline was then slackened and the wraps tightened with the haulback line and spiked. That line was then finished, and the haulback cut loose from the guy line and run in, as was the strawline.

Loading crew at Valsetz in the 1940s.Hayrack boom, pivoting on spar tree unshown at left, lifts logs from landing to truck on a plank road. The truck's tires and load are barely visible at extreme right. Charley Ames is the second man from left on the four-man log. His brother, Bud Ames, is the back man on the center log. Others are unidentified.   Photo from the collection of Charles V. Ames.

The raised spar tree was now held by three guy lines. The main line was slackened and slack pulled on it with the strawline so it could be taken loose from its stump and run into the yarder. The small block is taken in also. The tree was now ready for the climber to hang the rest of the tree plates and guy lines on it so they could be tightened.

Riggers working for South Coast Lumber Co. of Brookings, Oregon, had to raise a tree atop a knoll up the Chetco River. There was no lift, nor anyplace to get any lift. Barney Klahn described how "a little horse sense and ingenuity did the trick."

Raising a Spar Tree with an "A" Frame

"Whitie" Dowden was hooking on this job. He had the crew take two small logs, maybe 30 ft. long, and lash them together near the top with a chunk of strawline, making an A-frame. Straddling the top of the spar tree with the A-frame, and the bottoms of the legs of the A-frame pointing towards the butt of the tree, the top of the A-frame was raised up off the ground about15 feet with, I believe, the haulback line. The main line was run over the top of the A-frame crotch and back down to the top of the spar tree and shackled to it. By pulling on the A-frame and the tree, the A-frame gave enough lift to get the top of the spar tree off the ground. They both came up together. When the A-frame was straight up, we unhooked from it, and it fell to the ground.

With just that much lift, the spar tree came up easily the rest of the way. When you have no hills within a few hundred feet of the tree to be raised, you can hang the lead block in a standing tree near by and also use it for the tail hold. You hang at least two guy lines on the back side of that tree to keep from pulling it over. If you have neither a hill nor a standing tree for lift, you can raise a short tree probably at least 200 feet from the spar tree and use it for lift.

At Ozette Timber Co. on the Olympic Peninsula about 1941, I helped raise a spruce tree where I heard it said that this was the seventh time this tree was being raised. It could have been true. There wasn't a patch of bark left on that tree as big as my hand. Raising and lowering a tree that many times and moving it from setting to setting on railroad skeleton cars would surely take all the bark off.

There was an old retired hooker on that crew. He notched all the guy line stumps. He also picked them, too. No chain saws then, only axes, and he kept his as sharp as a razor. I needed an ax for something and I grabbed his as it was handy. Boy, was that a mistake! I thought that old man was going to eat me alive. Talk about getting raked over the coals. Nobody but him touched that ax. This kid learned a lesson real quick.

If you didn't run into any major problems, you could raise a tree, hang and tighten all of the guy lines, hang the blocks and boom, if you are loading with a boom, in from 1 1/2 to 2 1/2 days. If Murphy's law prevails, oh-boy, oh-boy, have fun."

Raising a spar tree with a Cat

When I worked for the California Barrel Co., near the little town of Orick on the northwest California coast, they raised spar trees with two Caterpillar® tractors. On one Cat, a 2U D8, they put all the line they could on the drum, over 200 feet. The other Cat, a 3T D7, had a regular winch line. The D8 Cat raised the tree. The D7 Cat could get around to most of the stumps to tighten all the guy lines. With Cats, raising a tree was a lot less work and speedier, too. There was no yarder to be tied down, fewer lead blocks to hang, and lots less strawline to pull.

The boss was a man from the Skagit River country in Washington. Whenever a tree was raised, after all the raising rigging and guy lines were in place, he would call all the men to come to the landing for a cup of coffee and a smoke. Then he took a tour around to all of the guy line stumps to make sure everything was as it should be. When everything was to his satisfaction and the tree started raising, very seldom did it have to stop before it reached the point where the first guy line had to be spiked. Taking precautions and having everything right paid off.

Taking a spar tree down

There is only one trick to letting one down: Don't get hurt. . . All of the rigging is taken out except three top or buckle guy lines left to hold the tree up. Two or three men take a railroad spike pulling bar and a hammer opposite the side where you want the tree to fall. They pull the spikes from the third and second wraps, take the wraps off the stump, pull the slack towards the spar tree, take a few spikes out of the first wrap, leaving enough to hold the guy line from slipping. This is where judgment comes into play. The bite of the guy line is taken over to the top of the stump and laid on top of itself where the first wrap starts. This is so when the guy line lets loose it is not pulled around the stump. The whip end of the line could hurt someone. This way it just drags towards the tree.

Next, take the top end of the spike bar laying it on top of the heads of the bottom row of spikes still holding the guy line. Pound them down with the hammer. At some point, the pull of the guy line will tear itself loose from the remaining spikes and the men step back and let her go. It is a simple process but you have to be careful. The pull of the other two guy lines will drag the tree in their direction. Jerk the remaining spikes out of the way, and you're on your way.

Topping the Spar Tree, the High Climber's Work

Bill Ames of Falls City, Oregon, was one of five brothers who were famous loggers in the 1940s and 1950s. Writing in 1987 for the Itemizer-Observer, he described the high-climber, the man topped and rigged spar trees, as "a special breed, with more guts than a pot-bellied cow" He was the highest-paid man on the rigging crew and the hardest man to replace. The following is blended from Ames two-part story.

The climber required a lot of nerve and stamina to start at the bottom of a huge tree and climb nearly to the top, cutting the limbs off as he went. After he arrived at a certain height, he had to cut off the top. This was like falling a tree, as there was still about 30 or 40 feet of tree left above him.

Topping the tree was the easiest part of the job for a climber, Ames wrote. As a rule, he could take his time and pick the day that he felt like doing it.

A good climber was always in charge and seldom made a mistake. The very nature of his job demanded that every move be made to count. Only two times did I ever see a climber drop a tool. Luckily, no one got hit either time.

The way climbers moved around up in the spar trees was a sight to behold. Even though they were tied to the spar tree with a rope, they moved around with all the aplomb of a chipmunk gathering nuts for winter.

A climber carried his tools with him when called and always arrived packing a pair of spurs and a belt. The spurs were the same as seen on the men climbing telephone or power poles, but they had longer spikes in order to penetrate the thick bark on the old growth trees. The belt was of leather about eight or nine inches wide with loops to carry tools sewn onto the belt. A special rope with an eye sliced in one end was attached halfway around it The rest of the rope was thrown around the tree and held in the eye with a cat's paw knot, allowing enough slack so loops could be thrown from right to left as the climber advanced up the tree.

A climbing rope was about an inch around and 30 feet long. It had a steel core so it couldn't be cut with an axe or pinched with a cable. It was a long way to the ground from up there and not very many ever survived a fall from a tall spar tree but some climbers survived falls of 50 feet or more without getting busted up too bad., Ames reported.

The greatest danger in topping a tree was in not cutting a deep enough undercut or not side-notching properly. This could cause the tree trunk to split, opening up like a book.

The climber, tied with a rope around the tree, would get squeezed to death. This did happen several times.

(Also), the climber was at the mercy of others when hanging the rigging. A wrong signal, or a failure of the machine below could be a disaster for the man working up there among all that mess of blocks and lines. Climbers tolerated no nonsense while up a tree.

Ames told of a "monster tree" at Bradley Woodard Logging Co., out of Westport, Oregon, a huge spruce about nine feet through at the bottom. "About 15 feet up there was a big growth on one side as big as a Volkswagen beetle. Ten feet farther up on the other side was another one just as big."

The regular climber took a look at it, and went home "sick".

Next morning, a new climber hired on, looked the tree over from top to bottom and walked off without a word.

Next day the hooker tried to do the job. He put on the climbing belt and the 30-foot rope was way too short, so he tied the pass rope onto the end of it. He came to the first big bump, realized it wasn't as simple as he'd thought, and called for Bill to come and give him a hand.

"Well, I cut a long pole and tried to poke his rope over that big bump but that wasn't going to work. If he made it over the first bump he still had more to contend with. At noon we watched him struggling to get over that big burl. After four hours he was still in the same place.

"His strength was gone. He climbed down, informed me that it was impossible for anyone to climb that tree, and collapsed. It was an hour before he got up and staggered away.

"My brother, Charley, showed up looking for work. I told the superintendent that if anyone could climb that tree, he could. The boss brought him out to the tree and showed him the climbing outfit. He didn't think that a young man like Charley could do it if three old hands couldn't. Charley was only about 20 years old. I watched anxiously as he prepared to go up that ugly tree.

"When he came to that first bump, it didn't even slow him down. He went around and around the tree, spiraling it all the time and two hours later was back on the ground.

"The regular climber made a marvelous recovery when he found out that the tree was topped. He looked very sheepish indeed when he found out a kid had topped a tree he was unable to climb."

For a photo of Bill Ames's brother, Charles V. "Charley" Ames topping a tree in his youth, please turn back to Chapter 3, page 52.

A Ledgerwood loader with metal tower on rails used a heel boom crane to load a train, 1940s. Photo courtesy of Port Gamble Historical Museum.

CHAPTER SIX

FIREWATCHING, EARTHQUAKES,
and FOREST MANAGEMENT

The Fire Watchman

In May of 1967 I retired from working at cutting timber. For five years I had worked on the cutting crew for Pope and Talbot Company at Oakridge, Oregon. With the start of the fire season in June, I took the job of fire watchman for the CR&K Logging Company.

This work required that I start work at one o'clock in the afternoon and stay at the landing until seven o'clock in the evening. At that time the watchman had to check the humidity gauge and when it dropped to 30 per cent, the crew would have to stop working. Later the rule was changed. The watchman had only to stay at the landing for three hours after the crew stopped working.

Then again, the forestry service made the rule that, as the crews went to work in the morning, they would see a notice posted as to the hour that the humidity would drop to the time when work must stop. Some days that would be two o'clock in the afternoon, or as early as eleven in the morning. On extremely dry days, no work was permitted. This was much the best system. The ruling covered the whole forestry district and relieved the individual crews of determining when they should stop work.

The summer of 1975, the CR&K Company enlarged their logging operations. They had five crews working in five different areas. This required five watchmen. They decided that a watchman should be at the landing at each of these areas, or sides, as they are called in logger's language. This meant that each watchman would need a tent, camper or trailer house and live on the job site. It was quite a problem to find watchmen who could or would stay out in a remote place every day for the duration of the dry season. The CR&K Company always expected me to help find watchmen and still watch one landing.

I liked this summer job, especially if we were logging a unit with a good trout stream so far back in the timber that it had never or seldom been seen by a fisherman. The reason the full-time watch-men were needed was due to the increase in theft and vandalism.

The previous summer someone had tried to drive off with a water truck, lost control coming down the mountain, and demolished the tanker when it left the logging road and crashed into the canyon. Then one of the loggers in the deer hunting season reported to the state police that they had noticed a car with out-of-state license leaving a logging sight. When this car was stopped by the police, they found it really loaded. It carried a 50-gallon drum full of gasoline, a power saw, several choker cables, and other smaller items they had picked up from various logging sites.

It fell my lot to be the watchman up on Fall Creek Mountain at an elevation of about 4,500 feet. This was how I happened to start writing up some of the events of my experience. Who better could write up a happening than the person who could say "I was there."

The Tillamook Burn in 1945. Three fires devastated this area in 1933, 1939, and 1945. Forest roads and watchtowers would have helped prevent such terrible devastation. More fire fighters could get in to areas jeopardized by advancing fire. Roads would also create barrier strip to slow spread of ground fire. Photograph by the author.

Earthquakes in Oregon

For years the people living in the Northwest thought that we were living in earthquake-free land. Then, on May 18, 1980, one of our mountains, St. Helens, blew its top and we started having disturbing experiences. In 1993, after Oregon had several severe earthquakes, we began to realize that this section of our country has many earth fractures. These fractures, after being dormant for many years, have now become active. When I look back on the many years that I worked at cutting our forests and building roads, I remember having encountered a number of these earth fractures. I have observed many earth movements that must have been due in part to these fractures.

When we lived in Tillamook County we knew of one earth fracture that could be traced from the Nehalem Bay across country to the Wilson River. It caused the roadbed to sink in one place. In other places, a hillside would move and close the highway. A prospector who traced this earth fracture found one place where blue mud came up to the surface in his back yard. With his gold pan he could recover tiny round pellets of mercury. At a place between the Kilchis River and Wilson River traces of gold were also found.

With our modern road building equipment we can literally move mountains and fill valleys. This often causes a disturbance and can cause a mountain to come sliding into our freshly made road. We often have problems in road building due to our slicing across a mountainside and crossing one of the earth fractures. When we cut the trees on the hills on the south side of Siletz Bay we found two of these earth fractures on a gash across the mountainside. One was an unstable area on a hillside with a steep slope above the bay. We thought little of this when we first cut and logged the area, but we ran into trouble when we started to build a road across this unstable area. We had noticed that, in a small canyon down near the bay, the land had slid off to bedrock.

The author stands by a large log he bucked at Jones Logging Company, working out of Lincoln City, Oregon, in 1959.

"Without this hard metal hat, I wouldn't be alive today," he says. Hard hats came to the woods in 1950, saving many a logger from injury and even death from blows from "widow-makers" —dead limbs or snags knocked down by a tree the logger is felling. The round object hanging from Brunson's belt is a 50-foot steel retractable tape for measuring log lengths and marking where to cut. Photo from the author's collection.

The farmer between the mountain and the bay had a small dam to catch water for his use. When our tractor operator started his road building across this area, he cut the road down to a solid base and just pushed the loose surface over the side. He was fortunate to notice that a snag below where he was working started moving. He quickly reversed the tractor and got it back on solid ground and watched a large section of the hillside slide down the mountain. The farmer was very unhappy when the slide wiped out his water supply and covered some of his pasture land. We had to abandon the road building on that side of the mountain and find a different way to get to the next landing.

I contracted to cut about a mile of right of way on Euchre Mountain in the Siletz Bay area. We had to go through a stand of old growth trees to log a blow-down on the south side of the mountain. We could tell that this was an unstable area by the dead trees on the mountain above the right of way. Sometimes a hillside will move just enough to break the roots of an old growth tree and kill it. Some of these dead trees had a downhill lean and we had to cut them, lest they fall while we were working beneath them.

The Tillamook State Forest seen from the Wilson River Highway in August, 1997. Hills that were stripped of life by forest fires in the 1930s and again in 1949 were restored to lush, green beauty by State reforestation programs, and nearly 50 years of growth. Photo courtesy of Paul Brunson.

Of the green trees, we needed to decide which to cut and which were safe to leave above the roadway. There were two trees standing about a hundred feet apart. One stood straight and we decided was it safe to leave standing.

On a Friday, the tractor operator made a road past this place. He made a cut in the hillside, about 15 feet deep. We were surprised on Monday morning to find that the stump of the tree that we had cut had split in two. Half of it was still in place and the other half had slid down into the road. Then, here was the tree we had saved, an old growth tree of 400 years and five feet in diameter, leaning across the road and split right up the center for a distance of near 40 feet! The split in this tree was wide enough that my falling partner and I could reach through and shake hands. We had to fall this tree and watch it break with the loss of that first-grade peeler log. This was only one of many earth movements that I have found in my working at logging.

It can be startling when you make a cut across a hillside, then see the whole hillside come sliding down, filling the roadbed you have just made. In 1923, I was cutting right-of-way up the Clackamas River when I first observed this to happen. About 30 acres of a heavily wooded mountain side came sliding down across the newly built roadbed and out into the river, partly damming the river. We had to go back and cut this tumbled mass of trees into logs. They had to bring in the yarder crew with a steam yarder and log off this slide before rebuilding the road.

If I were to guess where the next earthquake were to take place here in Oregon, I would pick the Oakridge area. A major fault makes a near half-circle there. It can be traced by earth fractures and at least four hot springs. It starts on Sinker Mountain between Fall Creek and Christy Creek and runs through the North Fork of the Willamette River, Salmon Creek, Salt Creek and Hills Creek.

A road builder out of West Fir had a big earthquake-caused problem. A whole mountain side moved and carried some of his equipment down the mountain side and buried it in the canyon too deep to be recovered for scrap iron.

One of my friends whose house was on the hill south of West Fir had trouble with his wells. He drilled a well that produced a good supply of water for his home. But, in 1971, it suddenly quit. Upon investigation, he found that both the water pipe and the casing were broken. We thought the break was caused by a ground shift. He brought in a well driller and had a new well drilled. His new well found no water. The ground shift had diverted the water source so that he no longer had water beneath his property.

I learned that a farm near Buena Vista in Polk County also had the experience of losing its wells. This farm had two good wells, but after this very light quake they could find no water under their farm.

There is no doubt that many ground fractures laying dormant now can become active at any time. The earthquake that struck the Molalla and Mt. Angel area had long been dormant. The quake in the Klamath area could be the result of pumping the hot water out of the ground for heating purposes.

Twin Rocks Creek

This stream flows into the Pacific Ocean just south of Twin Rocks. At about the time the railroad was built, this stream flowed into the north end of Smith Lake and outlet into Tillamook Bay at Barview. The real estate promoters in that day did things that now are no longer permitted. That outlet stream interfered with the buildings planned at Barview. It took only a shallow ditch to divert the stream from flowing into the lake and make it flow directly west across the sand dunes into the ocean.

Smith Lake with its sand bottom was one of our favorite places to swim when we lived at Rockaway. In the 1940s a logging company tried to get a permit to use this lake as a log dump. The whole community protested, especially the Methodist Church that had plans for a building on property of theirs between the lake and the ocean. We were thankful that the permit was not given.

When the north jetty at Barview was built about 1930 a spur railroad was built up Twin Rocks Creek to a rock quarry a little to the south side of that stream. It was about a mile up to this quarry. After the north jetty was built, a logging company used this spur to get into their first landing to log a tract of timber that belonged to one of the paper mills at Oregon City. This was a poor time to start a logging venture. Lumber prices and demand were so low that very few logging camps were operating. This company had a good reputation. They brought in more rails and a small locomotive to handle the loaded cars of logs. I went to work on the cutting crew, knowing that it would be more than a month until the first payday.

An old steam donkey was brought in for yarding. Some logs were sold to the mill at Garibaldi. Logs shipped on the railroad to a dump in the Willamette River went to the best market, especially for cedar logs. The logs suitable for pulp had to go to the paper mill at Oregon City.

In those days, the owner of the land that the logs came off had first claim on any money that the logs sold for. Some of us worked more than two months without a pay day. I don't know just who started the legal action, but the logging was shut down and a court appointed person put in charge. A large raft of cedar logs that should have helped pay the crew was towed down river and across the Columbia River. It sold for just the transportation cost. Here I was with a large bill at the grocery store and no money to pay. I was fortunate in that I got a load of lumber from the Garibaldi Mill, charged to the logs sold at that place.

I now could add onto the family house we were building. There were cedar slabs at the landing for roofing and siding and the hardware store gave me the nails on credit.

The author, on log, instructs son Paul Brunson on use of the power saw for log bucking. Photo from the author's collection.

We cut one of the largest trees near the landing on this creek. It had a dead top, but we still got three 32-foot logs. Someone asked me how long did it take two men to fall that size a tree. Well, it took more time than usual. We came up to the tree on Monday afternoon. We put in the spring boards and went up two boards high to get away from the root swell. On Tuesday we spent four hours, the two of us just chopping in the under cut that was 11 feet across. Even with this high stump, it was more than 12 feet in diameter at its widest place. We spent the whole afternoon sawing and wedging. We had to cut side notches so that we could cut it with a 10-foot falling saw. That took a lot of chopping on each side of the tree. We had only one place to lay that tree and save all the good wood. We were still sawing and wedging at quitting time on Tuesday.

Wednesday it stormed so that we didn't work. On Thursday we got it sawed through but we didn't have enough wedges to tip it off the stump.

That evening we went to the shop and found some heavy steel plates that were nearly an inch thick. With this extra wedging, on Friday we finally got that tree on the ground and into saw logs. They had trouble with the large butt log. They moved it to the landing, one end at a time. At the landing they found that it was too large to load on the railroad and at that time there was no log truck able to haul a log of that size. It could have been split into four pieces and shipped that way but, with the legal problems, nothing was done. That clear grained, 32-foot log was left to rot. In due time the equipment was moved out and the rails taken up.

I expect the stream is still a good trout stream. One good thing: The logging opened up the stream for fishing. There were large salmon trout that found their way in from the ocean. We often caught 12-inch trout in that small stream. Many changes have taken place in that country and along that stretch of beach land since the railroad came in, and the paved road was constructed from Tillamook to Nehalem Bay.

The railroad was built in 1911. The road, now US 101 and paved, first went as far as Rockaway Beach in 1926. It was a plank road. At that time it dead-ended at Manhattan Beach, now the northern part of the town of Rockaway.

Nelscott

This community is now the middle part of Lincoln City. The two men who plotted out the town site gave the town its name.

My wife's father, A. C. Stanbrough, and Mr. Nelson were friends of long standing. When the Stanbroughs decided to establish a summer home at the beach they bought a lot from the Nelson Scott promoters and built their vacation home.

Mr. Stanbrough was a skilled carpenter, as well as a college

The author and his son, Howard Virgil Brunson, falling a tree with a 50-pound 5-hp Mall power saw in 1946, cutting on a slant and using wedges to prevent binding of the blade in the cut.
Photo from the author's collection.

professor at Oregon Normal School (now Western Oregon State College) at Monmouth. He built several houses at times when school was out for the summer. I don't doubt that when the church at Nelscott was built he helped. He always willingly lent a hand to any community project.

When my work as a contractor for cutting trees into logs brought us to live in Lincoln City, it was only natural that we attended the church at Nelscott. We were welcome members to this church. My wife and I were always active church members and our two teenage girls were regulars in the younger age group.

We found this church had two problems. The first problem was the uncomfortable benches we had to sit in. It was a real torture for me to sit for the one-hour morning church service. One line of thinking is that if the pews are too comfortable some members will fall asleep. I know this from the experience of a person who is used to outdoor activities. When a person comes into a building to a seat that is warm and comfortable, it is hard to stay awake. One time when our pastor caught me drowsing off during his sermon, I thought it proper for me to apologize. He just smiled and said, "I thought you were nodding in approval." Now, most of us think that a comfortable pew will encourage church attendance. I know that when we move into a new community, a comfortable seat may help in our choosing the church we join.

The other problem was the clearing of a recreational area that the church owned. This tract of land had been covered by a stand of spruce trees. They had let a logger come in without a clear agreement on just what should be done. This logger worked with the purpose of making money. He cut and took out just the trees that made mill logs and left the trees standing that were only good for chipping. When they asked him why he was leaving these trees and the top part of the trees, he said that he could not sell these cull logs at a price that would make him money. Winter storms blew down most of the trees he left standing. Now they wanted a large enough space for a baseball field and a tennis court, but that space was all cluttered with the wind-blown trees and tops of trees.

The clearing of this playground was a challenge to me. I grew up with the idea that a good church person was expected to give time to this kind of a project. I was asked to take charge.

With two members of my cutting crew, I went in and felled the trees left standing. We cut logs of the wind-felled trees, and made logs out of the good parts of the trees left on the ground. I contracted a logger who had a truck and small tractor. He specialized in salvaging these poor grade logs, cut into 16-foot lengths, trucking them to a paper mill in Oregon City. He agreed to haul these logs off the ground if we gave him the logs with one stipulation: That we provide one man to help in the loading of the logs onto his truck. It ended up with me having to help loading in the morning and again about noon.

After we got the land cleared of logs, there was the problem of clearing the stumps off the ground. I took the problem to our logging superintendent. I knew that they had one of their large tractors in the shop for repairs. This man was also of the thinking that this land clearing was a worthwhile community project. He said that it would be a chance to test the repairs of that tractor. The company would clear the playing field if we paid the operator's wages. Well, I like to see a job well done and we found the means to do it.

Here the church got their land cleared at no cost to them, but I had trouble with the church treasurer. It seemed that the first logger had been paid $5 a load for each load of logs that he had hauled off that tract of land. We had a problem persuading him that we could not pay that fee for the truck loads of cull logs of 16-foot length. The other logger hauled 40-foot logs.

About this time I injured my left hip joint and spent time in the hospital. My wife and daughters started looking for a church with more comfortable pews. I could no longer endure the torture of sitting in the pews at the Nelscott church. It was the more comfortable seats that caused us to change our place of worship.

Problems Harvesting Our Forest Lands

There are several different kinds of rot that infect our growing forest. Some of these cannot be detected until the tree is felled. This is especially true of the trees in the Cascade mountains. Quite often in some areas, one half of the logs were left on the ground. When the demand for wood chips made it possible to salvage these logs, we went back and salvaged the cull logs. Now where the lay of the land permits it is common practice to go in and take out the known defective trees.

Wind storms are another menace to our forests. We have to use care in disturbing natural resistance to wind. I have often seen a wind storm blow down remaining the trees after a tract was selectively logged.

I am anxious to check to see if the old-growth trees we videotaped in 1996 survived a big storm that hit this section of Oregon. One old tree stands near Valsetz, Oregon. When this area was first logged, trees with a dead top and those with stump rot were left. At a later date most of these trees were cut in a salvage operation. This tree stands at a turn of the road surrounded by the new growth of trees. This old giant of a tree is 34 feet in circumference at shoulder high. The base is just a shell about two feet in thickness. I have felled a great many trees with stump rot. I would judge from my experience that there are still at least 10,00 board feet of good mill wood in the center of the tree if it could be felled. If the wind has felled this tree, very few good logs will be salvaged.

Today's Forest Management

We who have spent our lives working in our forest lands have a sincere desire to see the best utilization of this natural resource. We know from experience that when a tree has reached maturity and is deteriorating from disease or insect attack, or has been fire killed, it is time to remove it and replant.

The practice of going into our forest land for only a quick profit can no longer be tolerated. Our timber lands are no different than our farm lands. We plant, let grow, then at the proper time we harvest, as with other crops.

Due to the development of a use for wood chips we now can use all of the tree. We no longer take only the soundest part of the tree and leave the balance on the ground to be burned so that the land is cleared for replanting. We no longer burn the natural accumulation of moss and leaves, but let them remain as ground cover, needed to aid in the growing of the newly planted trees. We can proudly say that by our continual renewal of our forest lands, we will never run out of trees.

Further, we make better use of each tree. In the old days of cutting trees by hand, we had to spring-board up, that is, to cut notches in the tree into which the ends of our springboards fit. We stood high above the ground to get above the swell of the tree's roots and cut a straight-grained log. We had to leave huge stumps. If we tried to cut close to the ground with our handsaws, our saws would bind in the cross-grained wood of the root swells. With power saws, we can cut our trees near the ground now. That cross-grained wood no longer goes to waste.

With our modern mobile, collapsible towers, we can quickly move from one location to another to yard logs onto landings for loading onto trucks. With our modern road building equipment we can built roads closer to each cutting "side", making for shorter yarding distances. This saves time and labor since we no longer need to leave standing a suitable tree for a spar tree, nor does a high-climber have to climb, limbing as he goes, and top this tree. Much rigging time and labor is saved, and a dangerous job is no longer needed.

Willamette Industries clearcut this land in 1987. A year later it had been replanted with hand-span high nursery seedlings. Photo taken by the author in 1997 shows the new forest after a decade of growth.

For a hundred years the large timber operators, after cutting out their forest lands in the eastern, central and southern parts of our country, came to the West Coast and started cutting our vast forests here. Now that the eastern, central and southern forest lands are again covered with a trees large enough for harvesting, these companies are moving their operations back to harvest this second crop of trees.

Many of those older timber operators who were only interested in a quick profit are no longer in business. The Longview Fibre Co. has often bought tracts of their logged-off lands and salvaged enough logs to equal the price they paid.

One of the best caretakers of our forest lands is the Boise Cascade Company[1]. They have their Oregon headquarters right here at Independence, near Monmouth where I live. This company also

---

[1]Editor's note: Boise Cascade plants over two million new trees each year. This company manages 2.4 million acres of forest lands in Oregon, Washington, Idaho, and Montana. They have milling operations in Maine, also. When they log an area they leave trees along all the waterways, and in some places even plant more vegetation to shade the water and keep it cool. They place large logs in streams to provide the right mix of pools and riffles for fish to enjoy. When they build roads they build them in ways that cause the least soil disturbance and erosion.

According to Boise Cascade Communications Department, they use nearly 100% of each tree they cut down. Bark becomes fuel for power plants. Bits of wood too small for lumber, wood chips, or pressed wood become "true grit" sold for garden mulch. Their paper mills have mostly phased out the use of elemental chlorine. They manufacture recyled-content paper products throughout the U.S. Water used in paper milling goes through very strict purification processes before being returned to streams. M.W

conserves timber by making newspaper of fibers from wastepaper that other wise would clog landfills. They, with several other farsighted companies, will always have harvestable trees to keep their mills in operation.

Another company that takes conservation seriously is Plum Creek Timber, headquartered in Seattle. Their policy is to leave a 200-foot buffer strip of standing timber along each side of streams in a 170,000-acre habitat conservation area, though Washington State law requires only a 40-foot green strip. Watershed analysis reveals how wide a strip of uncut trees is needed in each area. These green strips provide shade for the fish, and protect watersheds. Plum Creek also plants two trees for every one tree they cut down. They not only renew their forests, they are doubling the number of trees for the next generation.

Weyerhaeuser, headquartered in Federal Way, Washington, also tree-farms vast millions of acres of their own forest lands. They give careful attention to the best practices to insure that they have plenty of healthy, growing trees in all stages of growth for future generations to harvest and turn into the things all of us need, from toilet paper to houses.

Some of today's logging companies not only care for their own holdings but buy up the cut-over lands of those quick-profit operators of other years. Quite often, under the present demand for wood, they are able to salvage enough materials to equal the cost of this land to them.

At the present time we have a needless shortage of logs. A certain group of people have only the thought of saving our old growth forest, and even agitate against the cutting of any trees. They don't realize that a tree can die of old age or disease.

In some areas of our Cascade Mountains there are old growth trees from 400 to 600 years of age that are rotting faster than they are growing. It is well where an old growth tree is healthy to let it grow but where mature trees are deteriorating faster than they are growing we find it is best to go in and harvest them.

In a new growth of trees it is often the practice to do what we call thinning. If the trees are too close together, it is possible to cut out up to a third of them and thus give the remaining trees more room for growing. We use the wood from those trees cut in thinning. We are now seeing the cutting of small trees, with usable logs down to two inches in diameter.

Where the lay of the ground permits, we can cut out and market some trees each year and have trees to harvest year after year. With the proper management we will never run out of trees.

Recently a friend on a visit to Europe took a copy of my logging pictures with him. The people he showed the videotape to were surprised at our vast forests of large trees. In his country for a thousand or more years they have been replanting their trees and saving even the dead limbs for fuel.

They keep their forest lands like parks. The day is here when we, too, must practice the best of forest management. This means selective logging where possible. On steep, rough ground it is best to clear- cut and replant.

I note that here in Polk County, the larger tree farms make their second harvest on land where the trees are about 70 years old. This is better than cutting at a younger age. The old-growth trees were about 400 years old when cut in 1915. Now a second cutting has taken place, and yet another new growth of planted trees looks good for the future. We can hope that no fire gets loose and kills them. In our forest lands the age of the trees, as shown by the number of rings we see when we cut them, tells us how many years they have grown and escaped forest fire and the timber crews.

In my more than 40 years of working in our forest lands, I learned how to cut a section of trees and not leave the standing trees so that a windstorm could blow them down. We found it best if we could leave smaller trees of about the same size, with no trees standing out or taller. On a south slope, the smaller trees give the wind an upward lift as it starts up a mountain side. The direction of the wind, the slope of the ground and the lean of each tree should be noted when planning the cutting of a unit. Most of the trees will have a downhill lean.

Fire-killed trees like these of 1939 Tillamook Burn can be salvaged, their wood put to good use, and the forests re-planted.
Photographed in 1941 by the author.

Nowadays, it is the practice to cut an area, then leave an area. Usually this will naturally reseed the cut area, but we get the new trees started much quicker when we replant the cut-over ground with seedling trees.

Managing Our Public Forest Lands

This is a very grave problem today. We are seeing so much of our publicly owned land set aside for parks, wilderness areas, game preserves and numerous other purposes. Some restrictions are imposed on privately owned lands.

We needed to set aside some lands. It is the care of our publicly owned lands that is of most concern, especially the more remote areas that can be only enjoyed by a few sturdy persons with backpacks. Why not build roads into these areas so that more people can enjoy their beauty?

Also, by building roads into these remote parts of our land, we can better control fire damage there. Following a forest fire, would it not be better to be able to salvage the dead and dying trees than to waste this valuable asset? How long can we just let nature take its course? After a devastating fire it takes nature a long time to regrow trees.

Much of our privately owned land is in tree farms, managed by people who have spent their lifetimes working with trees and have a natural love for healthy, growing trees. Trees are a renewable resource.

We find that to get the best out of our forests on the steep land, we must clearcut in some places[2]. It is only on the gentler slopes that we can selectively log and take out only the dead and mature trees. Some times we can thin out the trees that are too close together. This gives the remaining trees more growing room. Steep and rough ground must be clearcut and reseeded with hand planted young trees.

We find that it is mainly in the first 15 years of a fir tree's life that the wild creatures use these areas. As the trees grow larger they crowd out the plants the animals feed on. While it is true some may find refuge in the nearby old growth trees, most of the animals and birds feed in the new growth. They will just slip ahead of a person walking through the unit and we only catch a brief glimpse of them.

As for the spotted owl that we've heard so much about, in all the pictures that I have seen of them they are in the younger trees. In the more than 40 years that I worked in our forest lands, I can remember seeing only one owl in a stand of old growth trees.

Uncontrolled fire, wind, storms and bugs are the worst enemies to our forest. There are three kinds of rot that also attack our trees. The one we call stump rot works from the ground. Another kills from the top down. The third, ring rot, works in the center section of the tree.

The proven way to best manage our forest lands is to alternate strips of the timber, creating open, sunlit spaces between timber strips. The sun helps prevent rot getting started in the trees, and the open spaces give animals places to live and browse or graze on the plants that grow in sunlight. The strips of

---

[2]These clearcuts soon grow bushes and grasses for deer to browse and berry bushes for bears to feast on. Being protected by law from hunters and trappers, the numbers of deer, bear and cougar in the Northwest have increased to where more and more of them are extending their ranges into suburban areas. Editor

standing trees alongside these open spaces provide roosting places for wild birds like the famous spotted owl, which feeds on small rodents that live only in these forest meadows.

Trees properly spaced for good growth will, in about 20 years, crowd out the kinds of vegetation that the wild creatures eat. As the food supply diminishes, an area supports less wildlife. In the Cascade mountains, in great sections of the forest we find the trees are deteriorating from rot and old age and should be harvested and new trees planted to replace them. Soon bushes and grasses will grow up to provide browse for deer and food and cover for some wild birds and rodents.

We can build roads into the remote areas and salvage the fire damaged trees, mature and dying trees and those with a bug infestation in which caterpillars eat the foliage and the tree dies. When those deteriorating and dead trees are not taken out, they present an opportunity for a forest fire to get started.

A single strike by lightening can ignite those tinder-dry dead trees. The fire will kill all the new growth, then eat into the growing green timber. In

Vast hillsides covered with charred tree skeletons after the Tillamook Burn fires of 1918, 1933, 1939 and 1949. After salvaging this timber, loggers, volunteers and foresters replanted with seedlings that have grown to a new forest of spruce and Douglas fir. Photographed by the author, 1939.

time, if we leave it to nature, other wild fires will devastate much of our valuable national forest and national park land. Access roads into forested areas would help firefighters get to work faster.

We were shocked in the summer of 1996, when an uncontrolled fire trapped and killed a number of firefighting crew. These people had been sent into a rough, roadless area to fight the fire with only hand tools. Had there been even a primitive road into this canyon, modern fire fighting equipment could have gotten in and stopped that fire which, with a change of wind, entrapped and killed the people.

We get reports that the fire that swept through Yellowstone National Park killed one-third of the trees in that place of natural beauty. Nature will try to replace the fire damage. The big question is how soon the next fire will sweep through the area.

Salvage Logging

In the 1920's the only practical way to move large logs was by water or railroad. This led to a very wasteful use of our forest lands. Only the best grades of logs were of sufficient value to be hauled to market. Logs with any defect such as rot or large limbs were left on the ground. Small trees and defective trees were usually not cut by the cutting crews. These trees left standing helped to reseed the logged-off lands.

The development of bulldozers, log trucks, modern road building equipment, and portable metal spar towers led to a better use of our trees. With the demand for wood chips, it became profitable to go in and harvest most of the cull logs. Little debris is left behind and there is no longer any need to burn over the logged-off area to clear it for replanting.

On a recent trip up the Luckiamute River basins, we noted that much of that area first logged had been relogged to salvage the defective trees and poor grade logs. Now the new growth of trees is being logged there. Most of this land is so steep it is not suitable for selective logging and must be clearcut, then replanted by hand with small nursery grown trees.

Regard for our rivers and streams

Loggers no longer abuse our rivers and streams as they did decades ago. We need to remember some of the ways that the logging industry mistreated our smaller rivers in the past. Driving logs down a river gouged out the gravel bars the fish needed for spawning and took out the big boulders that provided resting places for them as they traveled upstream.

Another bad practice was the building of splash dams that blocked the fish from their natural spawning places. In a large river, like the Willamette, it can often be a help to navigation and no damage to the fish to remove some gravel. In fact, it can make quite a nice pond, suited to some species of fish. I remember driving by one stream where removal of gravel helped the fish by making a resting pool in a swift-flowing stream. My son Paul, then six years old, on seeing the machine at work exclaimed, "There is another river digger-up!"

Robert Lambert felling an eight-foot thick spruce tree with the new, lightweight power saws in 1956 near Kernville, Oregon. Formerly, with handsaws or older power saws, felling a tree this big was a two-man job. Modern milling methods let the logger save timber by cutting close to the ground, even though the tree widens at the base where the roots flare out from the trunk.   *Photo from the author's collection.*

Measuring the diameter of a huge spruce log in the 1940s. Root flanges were trimmed from this butt log to make it legal size for trucking. In those days the trimmed wood went to waste. Now it can be chipped for pressed wood, as can tree branches and other unusable debris from logging.
Photo courtesy of Tacoma Public Library, Richards collection.

I remember that when I worked for Markham & Callow Logging Company in 1949, the North Fork of the Nehalem River had a splash dam. When a stream did not have enough water in it to float logs downstream, the logging operators built a dam of cribbed-up logs. The dam had a gate, so when enough water built up behind the dam they could open up the gate. Then all the logs they had put into the pond behind the dam would go splashing down the stream with the released water. This dam had not been used for its original purpose for at least 20 years. It was a sturdily built structure that was used in 1949 as a bridge for Markham and Callow's log truck road. They only had to bridge over about a 30-foot gap to make a strong but narrow bridge for their trucks. Only after a loaded log truck ran off the bridge did they add safeguards.

In building the truck road up a small side stream, they took gravel out of that small stream and did much damage to the natural spawning areas. Now you must get a permit before taking any gravel from an Oregon stream. Taking gravel from stream beds for road building is no longer permitted in our smaller streams.

The Big Luckiamute River is another example of a river that was abused in the past. Logging companies floated logs down the Big Luckiamute before the railroad was built in the early 1920's. The farmers along the river were glad to sell their timber at a cheap price as this helped to clear their land for farming and pasture. But, when the company took out the big boulders and gravel bars along the river to

Loggers now use portable metal towers mounted on wheels. Towers can be collapsed, moved to a new site and raised with lines and engines like the one behind this tower. Photo courtesy of Tacoma Public Library, Richards collection.

for spawning. Now such rivers are being restored. Boulders or big logs are being placed where shady pools will provide necessary resting places, and gravel bars created for spawning places. Fish populations will grow again, and these rivers again be good fishing waters.

When this railroad was built, many spur tracks were built to the logging camps of the companies logging their timber holdings in this area. When the Depression stopped the logging in 1930, most of those logging operations closed down and just left their equipment standing where it was. By the time the log market improved, most of that equipment was badly worn and of little value. By then, log trucks and improved logging machinery had been developed. The railroads had been phased out. Charley Ames tells how he and his brother, both retired loggers, earned extra money by searching out abandoned logging equipment and selling it for scrap iron.

Now the whole Luckiamute area is growing a new forest. Charley thinks that when this forest is harvested, more of that old, abandoned logging equipment will be found. He thinks at least one of those old steam-powered donkeys will be found right where it yarded its last log in 1930.

Today's loggers realize we must protect and help restore our streams. We must leave a clean stream when logging and must put back boulders that were removed long ago. We must leave buffer strips of trees along rivers and streams for shade and a natural, pleasant place for the fish to live in and reproduce. We must adapt our logging methods to the lay of the land and work with regard for the animals who live there, the fish in its streams, and the future needs of our country and others for our timber.

Sometimes selective logging is best. Sometimes clear-cutting and replanting are the best methods to insure that a new forest of trees properly spaced for their best growth will cover the land.

On extremely steep hillsides, it is not well nor even sound business to try to put in a logging road. On ground like this, helicopters can lift the logs from where they have been cut and carry them to more level ground where they can be loaded onto trucks.

With our present-day equipment, we are able to get more board-feet from our trees by cutting close to the ground, where a few decades ago we had to put in springboards many feet above the ground in order to fall a tree with stump rot or root flare.

Now that every bit of the tree can be used, we no longer must leave spindly trees that blow down in the next windstorm. Small pieces can be combined in the factories to make laminated beams. Trees not fit for lumber for the building trade can be made into wood chips for pressed wood.

These new methods of logging make better use of our timber and preserve more of the watersheds and the natural habitats of other creatures.

A view of new growth in the old Tillamook Burn area along the Wilson River in Oregon. Lightest trees are alders. They act as "nurse trees" sheltering seedling evergreens for their first tender years. Darkest trees are Douglas firs. Photo courtesy of Paul Brunson.

Log rafts in several stages of being made in a cove in British Columbia. Photo courtesy of Tacoma Public Library, Richards collection.

Seagoing tugboats move four log rafts from inland waters toward lumber mills along the Pacific coast. Such log tows were common sights in Oregon and Washington coastal waters in the 1940s and can still occasionally be seen in Puget Sound. Photo courtesy of Tacoma Public Library.

CHAPTER SEVEN

WILDLIFE ADVENTURES

We who work in our forest lands get a close-up view of the wild animals and birds that live in our forests. Deer come to investigate our work. Bears surprise us and are surprised by us. We see many kinds of wild birds.

I have always liked to hike into the wild country. I like to find deer, then see how close I can get to them. Sometimes before hunting season, I will go out to see where the deer are, and how best to see them first. We don't see very many bears in the wilds. These animals usually stay in the brush. When we do see one, the bear is always as surprised as we are. Most of the bears in western Oregon are black.

Grouse and Grouse Ladders

The blue grouse like to live in large old trees with lots of large limbs. They feed on the new buds. These large old limbs, after 500 or 600 years, will have moss two- to six-inches thick on them. Lots of limbs slope out and down in a winter storm. The blue grouse can find shelter near the body of the tree.

It is a very interesting time of year when the blue grouse are mating. The male will sit up near the top of the tree and give out his mating call, a hooting sound. These grouse are sometimes referred to as "hooters." The female will come and light on a lower branch. The male will drop down to sit at her level. She immediately flies up to a higher branch. They repeat this action until they reach his perch up near the top of the tree.

We always called those old trees with many large limbs "grouse ladders." We used to leave them for the grouse and to reseed cutover areas. But a lone tree, standing taller than surrounding plants, will draw lightening. The dry moss on those large limbs was like tinder. When a fire swept up one of those trees, the moss would be ignited. Because a fire creates its own wind, the flaming moss would be carried off by the wind and start new fires Then all that new growth of timber that had been seeded by the big, old tree would be destroyed. As a safety precaution against fires, we learned it is best to cut the big, limby trees, leaving no lone, tall trees to attract lightening

Another bird that we call a "native pheasant," though it is really another kind of grouse, likes to live in our forests. This bird in the mating season sits on a log and makes a drumming noise with his wings. We then call him a "drummer."

These birds all nest on the ground. In the nesting season we will often see the female leave her nest when we fell a tree. When she sees that all is clear, she will come back to sit on her eggs. She soon sees that we can be trusted not to drop a tree too near her nest. Then she will sit trustfully as the trees fall. She trusts us not to disturb her until her eggs are hatched and she can lead her chicks to a safer place.

Deer
April 17, 1997

Deer are very curious animals and when we are working they often will come each evening after the crew has left for home. They apparently come to check on the work done each day. When working as fire watchman, I have watched as the same deer came of an evening. They would walk within 10 feet of me sitting in my pickup.

When felling trees I have often watched deer come and feed on the moss that grows on the fir trees. The deer love this moss and seem to think we have done them a favor when we cut down a tree so they can reach moss on the higher branches.

Deer like to bed down on a ridge at the top of a hill where they have a good view and can see the approach of any danger to them. Often when we come to work we find that deer have been lying within a few feet of our tools. Sometimes in a winter snow storm these mosses will come off the tree and the deer will search in the snow for moss as the snow melts.

I have found only one fawn, a day-old youngster that would not move from where his mother had hidden it. It was so frightened that its heartbeats shook its whole body. This is one reason that we should not touch a young fawn. The shock could be fatal.

I love to go out into our forest and watch the wildlife. They seem to know when they are safe and when they are endangered.

Below: Two National Guardsmen, assigned and on their way to fight forest fire, offer a carrot from their lunch to a young doe after a Tillamook County, Oregon, forest fire in 1949s. Photo courtesy of Oregon Historical Society, Neg. #66238.

Bear Stories

I went to look over what had been one of my favorite hunting ridges in the Tillamook Burn in June, 1936. I had seen several deer when a noise on the next ridge attracted me. It sounded much like a young pig's squeal. I looked across, and there a small bear cub ran down the steep mountain side. Just as this small bear ran across a less steep place, the kind of slope the loggers call a "flat", a young fawn deer jumped up and ran after him, and over the hill out of sight.

It was rough going and it took me several minutes to get over to the other ridge, just above the place where the fawn had lain. As I slowly made my way around the mountain side, here came the mother deer. She stopped and sniffed around the place where the fawn had lain. I was within 25 feet of her, on the hill above. The wind blew up the mountain, blowing my "man scent" away from her. She just looked at me. I talked to her, trying to tell her that her baby had run down the hill after that young bear. Finally, she sniffed and got a smell of me. No doubt it was that "man smell" that caused her to run down the hill and out of my sight.

I wondered where the young animals had gone. As I started down the mountain, I came to the top of a large fir windfall. I walked down this tree until, near its up-turned roots, I found the mother bear feeding her youngster. The little fellow was hungry while the mother was searching for a rotten stump with ants or grubs in it. She had been gone long past dinner time. When I stepped off the log, the mother bear stood up and kept between me and her baby. In bear language she said, "Mr. Man, don't come any closer!" I tried to tell her that I would not harm her baby.

Nearby were several small fire-killed trees. The mother bear took her youngster over to one of those trees and tried to get it to climb to safety. She put her big right paw to his little butt and shoved him as far as she could reach. There was no bark on these fire-killed trees, and nothing for the little fellow to fasten his claws into, so down he slid. The mother took the youngster over to another tree and shoved him up again. Down he came again. Then the mother bear took her baby and ran over the hill and disappeared into the thick brush.

Many of the cedar trees in our coastal areas have a rot that leaves the lower part of the tree hollow. When there is an opening, this makes an ideal winter home for a bear. One tree near Lake Lytle, in Tillamook County, had the top broken out at a height of about 60 feet. You could walk right into this tree, look up, and see the blue sky. This 10-foot diameter tree was just a shell with plenty of room for a grown man to lie down inside. It must have been a similar tree that Joe Champion lived in the first winter he stayed in the Tillamook Bay area. Joe is credited with having been the first white man to live in the Tillamook area.

In February of 1933, Eccles McCaw and I were falling timber for a small logging company near Rockaway, Oregon. We found one of these hollow cedar trees. Signs indicated that this was the winter home of a bear. We remembered how two men cutting timber just a month before had cut a similar tree and found a rudely awakened and grumpy bear. They had to kill it with their axes. Not wanting to trust our axes, Eccles went to our friend Bob Hauk's house, a mile back, to borrow his deer rifle while I stood guard with my axe.

It didn't take long for Bob to come back. By the time we had the tree ready to fall, several more curious people were also present. Bob stood with his rifle at the ready, expecting any moment to see the bear emerging from the hole in the base of the tree. When the tree fell, we could look down the stump into the hollow. We saw a well-used compartment, but no bear. It was a warm day. The bear must have heard us as we cut the nearby trees and left home.

That February was unusually warm for winter time. I remember how I had changed to light underwear. Some of the old timers just sweated it out until cooler days came.

Ten years later, I did see a bear come out of the hole in the base of a cedar tree. I had a crew of two sets of fallers and three buckers cutting timber for the Menefeee Logging Company on Lost Creek on the Nehalem River above Foss, Oregon.

That was a cold March day, cold enough that the snow on the ground stayed frozen all day. When I came up to where George Oakland and Jack Ramsey were falling trees, they had just felled a fir tree and were preparing to cut a nearby cedar tree. I noticed a hole at the base of the cedar. Jack had to chop a springboard hole on his side of this tree. George got out his pocket whetstone and started whetting his axe while he waited for Jack to put in his springboard. The first thing I noticed was the expression on George's face. He was staring at the base of the cedar tree. There, to our surprise, was the black head of a bear coming out of that hole. There was small windfall by it, and as the bear came out, it had to climb over this windfall tree.

Jack was busy chopping the notch for his springboard. As the bear jumped over the windfall, it stepped on Jack's foot. Jack shouted, "Get off my foot, you black S.O.B.!"

The man and the bear stood eye-to-eye for a split second, then ran in opposite directions. I don't know which of the two was more surprised.

Jack came walking back to where George and I stood, feeling rather embarrassed at not having stood his ground and let the bear do all the running. Soon, however, his and George's nerves quieted down enough for them to start cutting the cedar tree. When it fell, there was a good-sized bear bed. In it was a small cub bear. I reached in and picked up the little fellow. He was not afraid and cuddled right up next to my body. It was so cold that soon the cub was shivering. I put him back in his nest. It would soon be quitting time and, after we left, his mother could come back and take care of him.

When we got down to the landing, we told the rigging crew about our bear experience. Bill Taylor decided he wanted to take that cub bear home for a pet for his children. We tried to discourage him but after we left for home, while his crew waited, he went up the hill to the stump and got the little fellow.

All went well for a time. His family liked the little cub and it loved to play with the children. As it grew older, it got rougher in the playing, until the children were getting deep scratches. Mrs. Taylor never had much love for this small animal. The end came one day when she came home and found the sitting room couch badly damaged. She laid down the law--the bear must go!

Now, it so happened that a lady in our town, in her younger days, had worked with a circus. She had some animal training experience. Marie Terhune was glad to take this growing animal. She soon taught him some tricks. She would walk down our streets of Rockaway with the young animal on a "bear leash." On occasion, she even took him on the bus to Tillamook. As the bear grew older and larger, though, he became too much for Marie. A service station at Valley Junction had several wild animals and birds. The owner was glad to add a bear to his display. With some reluctance, Marie gave up her pet.

It was in 1954 that I saw a brown bear. I had a crew cutting a unit near Sand Lake, Oregon. We came to a cedar tree that had a hole at its base. When we felled another tree, it brushed against this cedar tree and out ran a small brown bear. It ran across the hillside and under some windfall trees. I went over, thinking to get a closer look at the little brown fellow. When I got near, a large black bear climbed up on one of the windfall trees and started walking back and forth. She began telling me in bear language, "Let my child be."

Not many minutes before the little brown bear ran out of the hollow cedar, we had talked about the possibility that it could be a bear den. One of the men had wondered as to the color. I had said that it would be black as there were no brown bears in western Oregon. Now I knew better.

When I was a Cub Master, I loved to take the young boys on trips. In 1946 I had a four-wheel-drive Dodge. It was an army surplus weapons carrier. Those young Cub Scouts loved to ride in it. I took some of them up an old logging road to Lost Creek. We camped out overnight, getting heavy moss off the maple trees to use under our bed rolls. The next morning, we all went trout fishing. I rigged each boy up with a pole cut from the hazelnut bushes, with a short length of fishing line and a hook with a small piece of worm. Some of these boys got a big thrill out of catching their first trout.

I noticed two of the boys going up a small side creek. Soon, here they came, running back and shouting, "Mr. Brunson, we were chased by a bear!"

I said, "I see that you made it back, but where is the bear?"

That bear was more startled than the boys. It had run up the mountain and was out of sight.

My nephew, Walter Whitmore, and I had an interesting experience one morning while driving to work. The crummy with the rigging crew was just ahead of us. Walt and I followed far enough behind to keep out of the dust on that winding, graveled mountain road. I sometimes think that wild animals get a thrill out of dashing across the road in front of a vehicle. At other times, they wait and cross behind the moving car or pickup.

As we came around a bend in the road, a large black bear jumped out into the road. He started running ahead of our pickup. He was making great strides. Every time his hind claws touched the ground, he picked up a paw full of gravel from the road. The gravel came pelting back towards our windshield. One larger stone cracked it. That big black bear ran about a hundred feet up the road until he found a place to jump off into the brush.

While the rocks were pelting the windshield, Walter had ducked down. When all was quiet, he straightened up. He asked me, "Who will believe that a bear threw a rock and broke this car's windshield?"

Snag left as a perch for eagles and other wild birds rises above new growth of hand-planted, nursery-grown trees. Only in open areas like this can spotted owls and other predator birds find the rodents they feed on. Unfortunately, lone snags often attract lightning, resulting in wildfires that destroy these new forests.

Photo from the author's collection.

The Spotted Owl Controversy: It is for the Birds
Written March 5, 1996

The spotted owl made famous by the ecologists are rarely if ever seen in our old-growth forests. In nearly 50 years of work in the Coast Range forests from Cape Tillamook Head on the North coast of Oregon to as far South as Cape Mendocino, California, I can remember seeing but one owl in a forest of old growth fir trees. Some people give the birds first use of our forest lands. I doubt their claim that the spotted owl must have a stand of old growth fir trees in order to continue their existence. The only pictures we see of this owl are in trees not older than 100 years. It takes 300 to 600 years for a fir tree to reach its maturity and be classed as "old growth". They need younger forests where there are open spaces where the small rodents they feed upon can live.

Nor do any sea birds nest in old growth trees. The only place where we find the sea birds nesting is on the rock cliffs above the dashing waves. It is true that a great many species of our birds use a man-made nest if it is build so that it meets their nesting needs, but those are land birds.

We have tried in selective logging to leave some dead trees that are especially of the type our wild birds can use for resting or nesting. Often these snags attracted lightening that started fires. How can we protect both the birds and the forests?

It is during the first two decades after a clear-cut that our wild life gets the most use our of our forest lands. It takes the new growth about 20 years to shade out the grasses, berries, seed-bearing plants and the tender branches that provide food for our wild life. When trees grow close together and shut out the sunlight, little of food value can grow.

Our greatest dangers to our wildlife habitats are uncontrolled fire, old age, tree diseases and insect damage. To create and preserve the places where wildlife can live and flourish, we must find solutions to all these hazards to our forests, as well as managing our logging operations in a manner that regards the welfare of birds and animals.

Are Loggers Superstitious?

When asked this question, I can only reply: There is no place in logging for uncertainties. When I first started working in a logging camp, when I received instructions from the hook tender on how he wanted the work done, he asked me, "Now, do you understand?"

I rather indefinitely answered, "Guess so."

He very positively told me, "We can't have any guess work here."

He made sure that I did clearly understand his orders.

I found among the loggers I knew only one belief that might border on "superstition" and that is the belief in a guarding angel. The logging industry is rated very hazardous at its best. When I started working in our forest lands, in most of the larger operations the logs were hauled by railroads. The logging camps could only be reached by the companies' private railroads. Contrary to the preponderant belief of the workmen when in town spending their hard-earned money on a wild time, every well-organized logging camp had a group of steady workers. Most of them knew that they had to give a good day's work for their pay if the company was to stay in the business.

When in an emergency, I heard a workman calling on God for help, sometimes it was an earnest plea. I found that there were a lot of loggers like myself with very sincere beliefs in God and "guarding angels." How else could we survive so many near-fatal accidents?

At most of the larger logging camps there were houses for family men. Whenever there were enough children to justify a one-room school, and where there was a schoolhouse, there would be a group of people meeting on Sunday for at least a Sunday school for the children. My wife and I often conducted these Sunday worship services. I remember that when, in 1926, I worked at the Shevlin Hixon logging camp near La Pine, we held our Sunday School classes in the same building as the Saturday night dance.

I have never preached a sermon but, quite often, at church time people would come and expect a worship service. And there was no way out of taking charge. When a prayer was need, it often fell my lot to take charge of that.

Charley Ames as boss of Valley Lumber Co., 1956, with a three-log load on his Mack truck. Behind the truck is a Skagit loader with heel boom crane on wheels. Soft hat and rolled, not stagged-off, pants show that Ames was not on a cutting crew.

Photo courtesy of Charles V. Ames.

LOGGER LINGO

What the woods workers' words mean (some have various meanings):

Bight
A bend or bow in a line which can cost a logger's life if he's on the wrong side when the line straightens.

Bind of the Line
A cable put around the log so that by pulling on the top line the log can be rolled to position for loading.

Blowdown
An area of trees felled by a windstorm.

Boom
Logs used as part of floating corral for logs in a raft.

Yarding logs with a bulldozer capable of making its own road and lifting front ends of logs to avoid hang-ups on obstacles. Note the safety canopy over the driver. These men are logging a stand of Doublas fir timber. Photo courtesy of Tacoma Public Library, Richards collection.

Buck (verb)	To cut a felled tree into logs.
Bull block	The large block up on the spar tree. The main line goes from the yarder up through the bull block, then out to the head block.
Bullbuck	Boss of the cutting crew.
Butt	End of first log cut from tree, the end nearest the stump.
Butt rigging	Cables attached to the main line.
Caulks	Sharp, tapered, short nails that protrude from soles of loggers boots, especially boots worn by high climbers and cutting crew.
Chaser	Man working on the landing.
Choker	Short length of cable put around log to be yarded (hauled to cold deck).

Choker setter	Man who puts the choker cable around the logs.
Chute	Bare path down a very steep hillside down which logs are slid, usually into a pond or river.
Clawson	Device to hold knob on choker.
Clearcut	To cut all the standing timber on a site, leaving the land ready for new trees.
Cold deck	Logs hauled in from where they were cut to a temporary storage place.
Corduroy road	Skid road made by laying small logs or poles horizontally.
Crotch line loading	Hooks are suspended from the main line which runs from the head block back to tongs set in each end of the line. With pressure on the haulback, we tight-line the log onto a truck. A "poor man's way of loading logs."
Crummy	Vehicle for transporting loggers to work.
Cutting crew	Fallers and buckers--the men who fell trees and those who cut them into logs.
Donkey	Engine that powers yarding of bucked logs.
Donkey puncher	Operator of donkey engine.
Faller	Man who cuts down trees.
Flume	Steep wooden, water-filled trough down which logs were sent from a hilltop landing down to a millpond, or boards rough cut at an uphill mill to a finishing mill at a lower level on a road or railroad.
Haulback line	Goes around area to be logged, through side blocks, head block and back to be connected with main line.
Hayrack boom	Rack made by tying two poles with cross-pieces. Hooks or tongs are suspended from the rack for moving logs from loading site to truck or train.
Head block	The block on the spar tree to which the main line runs from the yarder.
Heel boom	A rack against which the butt or 'heel' of a log rests as it is lifted and moved from landing to log truck.
High climber	Man who limbs and tops trees to be rigged as spar poles or spar trees.

There she goes, down the hill! The final cut made, a Douglas fir falls. Photo from the author's collection

Hoot-owling: Working very early in the morning before humidity drops to dangerously low levels that would increase likelihood of forest fire.

Knob: On end of choker cable--knob goes into notch on butt rigging.

Landing: Where logs are stored until loaded on railcars or trucks.

Main line: The cable used to pull the logs to the cold deck, or landing.

Parbuckle: Way of loading by pulling on a chain around and at right angles to log.

Pass line: The first line the high climber fastens onto the top of a spar tree; a pass line pulls up the heavier rigging lines and bull block. Like a straw line.

Peavey: Tool with a movable hook on the end of a sharp pole, used for turning logs or levering them into position. Named for the inventor, J. Peavey.

Peeler log: Log suitable for rotating against a saw at a mill to make veneer.

Rolling Hitch: Knot in the cable around a lock, set so that the log will roll as it is pulled forward, thus moving away from whatever is in its way, such as a stump or another log.

Shackle   A U-shaped device to connect two ends of a cable with eyes on each end, also used to fasten a block in place.

Show   Any logging operation.

Side   Small cutting area within a show.

Side blocks   Blocks that keep the haul-back line in place.

Sidewinder   Limb or tree hit and knocked down by a felled tree as it falls. A sidewinder, may fall anywhere, may injure a logger.

Siwash   Stump the straw line is wrapped around, usually used to pull blocks out to the stump when blocks or head blocks are used in logging.

Spar tree	Tall tree saved, or erected, where lines attached to it bring in logs from where they were cut to a landing.
Strap	Short cable to go around a stump--has an eye on each end.
Strawline pulled	Thinner line used to pull out the haul-back line. A strawline is usually out by hand. It has numerous uses.
Tailhold	Also pronounced "tailholt". A single line to a solid object such as a tree or stump for a brake or safety line.
Tail tree	Tree to which the tail hold line is attached to keep a donkey from sliding
Tight line	Signal to donkey puncher when we have a log or other object suspended between two lines, the tailhold and main lines, by cables attached at both ends of the object. As the tailhold and main lines tighten, the cables lift the object.
Timber bind	When a saw sticks in the cut due to cross-grain in the wood where a tree's roots flare out from the trunk
Tin pants	Heavy canvas pants the loggers have waterproofed with melted paraffin. Used especially on the west side of the Cascade Mountains before modern waterproof clothing was invented.
Undercut	To cut a wedge of wood from the tree to make it fall where the logger wills; Also, the cut itself.
Windfall	Tree knocked down by a high wind.
Whistle punk	Signal man, usually a beginning logger, who signals the donkey puncher by pulling a cord to sound a whistle to tell when to go ahead, slow, or stop.
Widow maker	A dangerous tree, such as a dead snag, that in falling can kill a logger.
Yarder	Machine that pulls logs attached to cables (lines) to cold deck or landing.

A faller knocks the wood out of the undercut he has power-sawed in a big cedar tree in the Oregon Coast Range. Photo courtesy of Tacoma Public Library, the Richards Collection.

Logger's Old Saws

1. Never try to buck a tree from the downhill side.
2. Don't let the donkey run away.
3. Always shout a warning "Timber!" or "Down the hill" before your tree falls.
4. Always have a safe escape route.
5. Never mind trying to save the saw, get yourself to safety.
6. Don't get yourself in a bind.

Watch for the next Brunson Watkins Book

I REMEMBER FAMILY HISTORY
by Howard Brunson.

You'll read about life on a Chehalem Mountain farm in the early years of the Twentieth Century, about homesteading, about the author's parents' constant struggle to raise eleven children on 80 acres, 40 of them arable.

You'll read a real-life prodigal-son story.

You'll learn how Mountain Top people grew hop vines, about the blend of hard work and companionship at harvest time, and how the hops were dried, sacked and taken to market. That was back when work was done by man, woman and child, and by horse power when that meant power supplied by real horses.

You'll smile at adventures with the first automobiles, when driving was for the hardy and bold spirited—and be glad you didn't have to patch inner tubes after a tire blow-out, crank the car to start the engine, drive without paved roads, in an open-air vehicle with no heater, radio or tape player

You'll read the love story of Howard and Olive, and of their elopement on a snowy Valentine's Day Eve.

The story of the journey and the treasure of the lost wagon train, part of the history of Oakridge, Oregon, will grip you.

You'll follow Howard and Olive Brunson from a hill farm on Chehalem Mountain to Coast Range logging camps, to his semi-retirement at age 79 to life as a Willamette Valley beekeeper, orchardist and lapidary, then as author of two books.

To order additional copies of

I Remember Logging

Please send _______ copies of *I Remember Logging* at $29.95 for
each book, plus $5.00 for postage and handling for the first book,
$2.00 for each additional book in the same order. Phone for quantity
discounts (206) 463-9626 or Fax (206) 463-6688.

Enclosed is my   check or   money order in the amount of $ _______
Visa   or   Mastercard Number _____________________Exp. _______

Name____

Ship to____

Address ____

City _______________________________State & Zip ____________

Fir Tree Press
8903 SW Bayview Drive
Vashon, WA 98070

··

To order additional copies of

I Remember Logging

Please send _______ copies of *I Remember Loggin g*at $29.95 for
each book, plus $5.00 for postage and handling for the first book,
$2.00 for each additional book in the same order. Phone for quantity
discounts (206) 463-9626 or Fax (206) 463-6688.

Enclosed is my   check or   money order in the amount of $ _______

Name____

Ship to____

Address ____

City _______________________________State & Zip ____________

Fir Tree Press
8903 SW Bayview Drive
Vashon, WA 98070